MW01487967

0 00 08 0184072 7

TRANSFORMING POWER OF TECHNOLOGY

TRANSFORMING POWER OF TECHNOLOGY

THE PERSONAL COMPUTER

Sandra Weber

CHELSEA HOUSE
PUBLISHERS
A Haights Cross Communications Company

Philadelphia

Frontis: Personal computers, like this Windows XP Media Center, enable individual users to perform a variety of tasks from basic word processing to accessing the Internet.

CHELSEA HOUSE PUBLISHERS

VP, NEW PRODUCT DEVELOPMENT Sally Cheney
DIRECTOR OF PRODUCTION Kim Shinners
CREATIVE MANAGER Takeshi Takahashi
MANUFACTURING MANAGER Diann Grasse

Staff for THE PERSONAL COMPUTER

EXECUTIVE EDITOR Lee Marcott
ASSISTANT EDITOR Kate Sullivan
PRODUCTION ASSISTANT Megan Emery
PICTURE RESEARCHER Amy Dunleavy
SERIES AND COVER DESIGNER Keith Trego
LAYOUT 21st Century Publishing and Communications Inc.

A Haights Cross Communications ◀ Company

http://www.chelseahouse.com

First Printing

1 3 5 7 9 8 6 4 2

Library of Congress Cataloging-in-Publication Data

Weber, Sandra, 1961–
    Personal computer / by Sandra Weber.
        p. cm. — (Transforming power of technology)
Includes index.
Summary: Discusses the effects of the invention of the personal computer on
society and everyday life.
    ISBN 0-7910-7450-1
    1. Microcomputers—Juvenile literature. 2. Computers and civilization—Juvenile
literature. [1. Microcomputers. 2. Computers and civilization.] I. Title. II. Series.
QA76.52.W43 2003
303.48'34—dc21

                                                                2003009474

A PC on Every Desk

On January 3, 1983, *Time* magazine published its annual "Man of the Year" issue for the previous year. The article declared that "the greatest influence for good or evil" was not Ronald Reagan, John Cougar, or Leonid Brezhnev. It was not even a man; it was a machine — the computer.

"In 1982 a cascade of computers beeped and blipped their way into the American office, the American school, the American home," the *Time* article explained. "The 'information revolution' that futurists have long predicted has arrived, bringing with it the promise of dramatic changes in the way people live and work, perhaps even in the way they think. America will never be the same."[1]

THE POWER TOOL

In the 1960s, computers were mammoth machines hidden in special, air-conditioned rooms with raised flooring. They had magical blinking lights, vacuum tubes, and their own strange language. Most people had never even touched a computer, let alone dreamed of programming them or having one on their desks.

The engineers, scientists, professors, and college students who used computers found them very useful for doing complicated calculations that would take months to do by hand. Some of these people loved to write computer programs, regardless of the task the program was to accomplish. They were fascinated by the machine and hated having to share it with others. They considered the brief stretches of computer time they were allotted to be far too short.

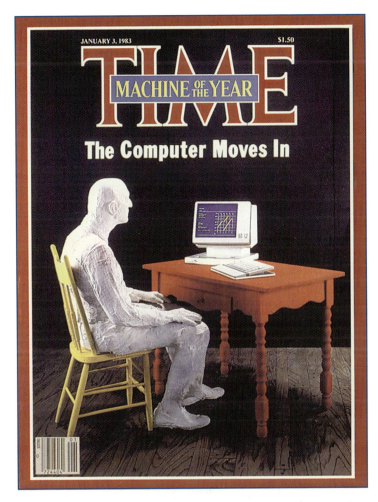

In *Time* magazine's annual "Man of the Year" issue for 1983, the magazine broke with tradition and declared not a *man* for 1982, but a machine — the personal computer. Doing so acknowledged the tremendous impact the PC (personal computer) was already having both in the workplace and at home. "America will never be the same," wrote *Time*.

As computer hardware components became smaller and less expensive, a few engineers realized that a computer could be made to fit on a desk. Computer programmers and

Paul Allen, at left, and Bill Gates started Microsoft in 1975. They vowed to have "a computer on every desk and in every home running Microsoft software." By 2003, more than 64 percent of homes had PCs, many of them running Microsoft's Windows platform.

enthusiasts, so-called nerds, imagined that they could have their very own computers. And, by the mid-1970s, they could—if they were willing to take a bag of parts and put one together themselves.

Bill Gates was an undergraduate student at Harvard University when Paul Allen ran into his dorm room with a magazine article about the Altair computer kit. "As the legend goes, the pair immediately recognized the Altair as a useless miracle. Due to the complexity of programming the device, it couldn't do much."[2]

However, Gates and Allen foresaw that these clunky boxes of electronics could do more—all that was missing was useful software. Within months, the two young men were selling the first software language program for that simple computer. They founded Microsoft Corporation and vowed to put "a computer on every desk and in every home running Microsoft software."[3]

Gates later reflected:

> Over the years, the PC [personal computer] has grown from a hobbyist's toy into an indispensable tool that continues to change the world . . . . And the little company Paul and I dreamed up sitting around my college dorm room is now the world's biggest software company, employing almost 40,000 people in more than 50 countries. From our roots in programming languages and operating systems, we've ventured into just about every kind of software you can imagine, from industrial-strength servers to games.[4]

In the over 20 years since it made its debut, the personal computer, or PC, has changed the way people communicate, shop, retrieve information, and entertain themselves. Like a Swiss Army knife, some of its attachments are more popular than others. It can be used to help with homework, balance a checkbook, solve a math problem, draw a cartoon, monitor investments, prepare tax returns, track family genealogy, and play games. It all depends on what software is running.

Initially, some people tried to dismiss the personal computer as a mere game or a toy. Others felt it was a glorified type-writer or adding machine. But the PC clearly achieved more when it replaced gears and levers with electronic circuitry. It put the power of computing into people's hands. And, although some feared that the computer would replace human workers, the computer industry has created jobs and given the U.S. economy a big boost.

Personal computers have been brought into offices, homes, and schools around the world. It is estimated that more than 500 million PCs are in use worldwide, a number equal to the number of cars on the road.

Personal computers are now commonplace in the United States. More than 64 percent of U.S. homes have a PC. More

Linux

The Windows operating systems set the industry standard and gave Microsoft a monopoly on the industry. There are few legitimate alternatives to Windows, but Linux is one of those brave few.

"There's no question that Linux is where it's at in computing today. Suddenly it's the hip new alternative, the free antidote to what Linux backers consider to be software's feudal ruling class, led by Bill Gates," wrote Doug Bartholomew in *Industry Week*.*

Thousands of fans of the Linux operating system gathered in California in early 2000 for Linux World. Many of them wore ponytails and dressed in cowboy hats, leather vests, and red top hats that said "LINUX" in tribute to the system's Finnish creator, Linus Torvalds.

Not all business executives are excited about the idea of running their company's computers with software that is free and has a penguin for a mascot. Slowly, though, corporations are installing Linux on their systems, especially as more software applications and consulting services support it.

When Torvalds was asked at the 2000 Linux World show if he was trying to capture a particular market with his operating system, the unassuming programmer responded, "I personally haven't tried to capture anybody at all."

Regarding the penguin logo, Torvalds said it was his wife's idea. "We wanted to find something friendly and not too serious. I was bitten by one in Australia once and I got penguin 'disease.' But we didn't test the penguin logo — it just works."**

* Doug Bartholomew, "Lord of the Penguins," *Industry Week* 249, no. 3 (February 7, 2000).

** Ibid.

than half of those PC owners say their computer is more important than their television; in fact, they call it the most important
device in their homes.

In the book *InfoCulture*, author Steven Lubar says that
people tend to attach too much importance to new machines,
greeting them with enthusiasm and investing them with their
hopes and dreams. The personal computer and the Internet
have been characterized as democratic technologies that give

**Linux, a company that provides free computer operating software,
is represented by this friendly penguin logo.**

the individual power and independence. However, these democratic hopes have dissolved into commercial products and uses. Lubar says, "The personal computer, which many had hoped would serve as a tool of personal liberation, found its first major use in manipulating financial information, and before long it was just another business tool." [5]

Was it really just another business tool? For some people with disabilities, the PC was the path into an amazing new world. Advances in computer hardware and software, combined with the Americans with Disabilities Act of 1990, eliminated some of the barriers to employment for disabled people. Almost 20,000 technology-related products are available for the disabled, including large-type screens, screen magnifiers, voice synthesizers, voice mail, sign language converters, and sip-and-puff devices that control other objects when attached to a PC.

"For the vast majority of disabled individuals, the micro-computer doesn't simply represent the ability to accomplish tasks a little faster or a little better. It represents the ability to do things previously considered unthinkable," claimed the annual *Computers in Society* in 1992.[6] This newfound ability helps the disabled achieve a new sense of personal identity and improves their self-esteem. The PC can change their lives.

In the business world, PCs became an even more important tool when they were networked together in the 1980s. Sharing information and resources among PCs made sense, and by the mid-1990s, PCs enabled the launching of the Internet revolution.

In his book *Nerds 2.0.1*, Stephen Segaller points out that, because the PC was developed after the Internet was, it really had nothing to do with the success of the worldwide network. Yet, the PC is the machine that brought the Internet into the business office and the home. Without the PC, "networking might have remained stuck in the limited enclaves of computer-science departments, federally funded research projects, and a few large corporate ventures." [7]

Today, almost 20,000 technology-related products are available to help the disabled, including screen magnifiers, voice synthesizers, and sip-and-puff devices like the one being used here with an early model Macintosh. Rather than using her hands to push the computer's controls, the user breathes into or sucks air from the tube to command the computer, which then controls other connected devices.

The PC spurred networking, incited the Internet craze, and more. In *Fire in the Valley,* authors Paul Freiberger and Michael Swaine proclaim that the PC "lit a fire that changed society." They further maintain that "the personal computer—and the Internet and other technologies that have been built on it—is changing the world as profoundly as the printing press or the Industrial Revolution. Yes, it's that big."[8]

HARDWARE AND SOFTWARE

What is a personal computer, or PC? It is a small, relatively low-cost computer that is sometimes called a desktop computer because it fits on a desk. Smaller versions that fit in a briefcase are called notebook or laptop computers. Because they are smaller in size than big mainframes or minicomputers, PCs are also referred to as microcomputers.

Like the powerful, room-sized computers that were their predecessors, PCs are general-purpose computers capable of being programmed for a variety of tasks. Programs and data are handled by the microprocessor—the brain—of a PC. It is only one of the physical components that make up the hardware.

Personal computers typically include several hardware pieces: a system unit, monitor, keyboard, mouse, and speakers. The system unit is usually a rectangular box containing the microprocessor, memory, disk drives, modem, and connecting circuitry. Disk drives might include a floppy disk drive, a hard disk drive, and a compact disc (CD) drive. The microprocessor, which might be a Pentium or PowerPC, sits on a single silicon chip the size of a fingernail. It is sometimes referred to as a "computer on a chip."

The keyboard is used for entering typographical characters. The monitor displays those letters and numbers, as well as graphics. The mouse allows the user to point at areas on the screen with a cursor. Speakers, a camera, a printer, a scanner, and other devices may be added to the system.

How does the PC work? Input is received from the user. Data and instructions are loaded into memory—specifically, random-access memory (RAM)—and held there until the processor accesses them. The processor executes the instructions, performing the necessary logical and arithmetic computations, and then puts the results in the memory. Output is presented on the monitor or can be sent to a printer or other device. The contents of RAM are erased when the PC is powered off, so in order for data to be retrieved at a later time, it must be stored on floppy disk, hard disk, tape, or CD.

A modem can be installed in a PC to connect it to the Internet over a telephone line. Alternate methods of accessing the Internet include cable modems, digital satellite services, or

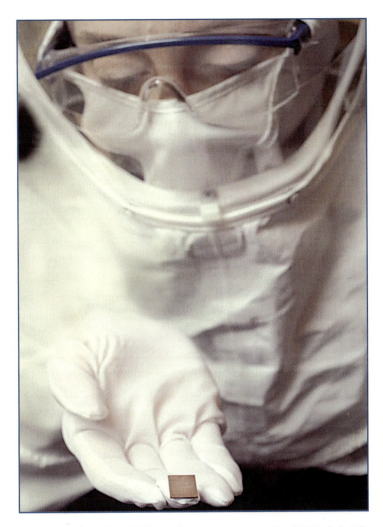

The computer's "brain" is a microprocessor, an integrated circuit mounted on a silicon chip the size of a fingernail. Microprocessors, like the one from Intel Corporation shown here, execute instructions that tell the computer what to do.

Wi-Fi networks, which use radio signals to communicate with a wireless router and wireless cards in a PC. Other kinds of circuit cards (also called accessory cards) will allow a PC to connect to devices such as a Local Area Network (LAN). This lets a network of PCs share resources, such as corporate sales data, spreadsheet and database software, printers, and disk drives.

When purchasing a PC, people usually focus on the hardware, but a PC needs software, too. It is actually the software that makes the hardware worth owning. Legend has it that many early buyers of Apple II computers walked into stores and instead of asking to see the Apple computer, they asked for "the VisiCalc machine." It was the VisiCalc spreadsheet program that sold the hardware.

PCs run several kinds of software, or programs. The operating system software is the basic set of instructions that underlies the computer system. It manages the internal computing and data exchanging activities and tells the disks, microprocessor, and other hardware what to do. An operating system, such as MS-DOS, Windows, UNIX, or Linux, is required for the PC to be able to run applications.

Application software allows the user to write letters, plan budgets, play chess, and perform other tasks on the computer. Most users buy "off-the-shelf"—packaged applications such as word processors, accounting programs, graphic design software, and games that are sold to the general public. Applications can be customized to meet the needs of a particular individual or company, but this requires the skills of a professional programmer.

Word processing packages help users assemble words and pictures to create party announcements, diaries, procedure manuals, essays, newsletters, reports, letters, and more. The appearance of documents can be enhanced with fonts, colors, and graphics. Two important features of a word processing package are the ability to save documents and to view them on the screen in a form called WYSIWYG (What You See Is What You Get). Other helpful features are spell check, cut-and-paste capabilities, and mail merge.

Desktop publishing software provides even more complicated layout capabilities than word processing programs, and these result in professional-looking documents. Instead of using typesetting machines, brochures, pamphlets, menus, and greeting cards can be created on a PC and printed on laser or inkjet printers.

Presentation software has replaced overhead projectors and carousel projectors in meetings and classes. It allows users to create slide shows or displays of images for group presentations. Charts, patterns, sounds, and video clips can also be included in the presentation.

A spreadsheet is a grid of rows and columns that can be used to represent accounting ledgers, teachers' grade books, home inventories, and budgets. With spreadsheet programs, it is easy to enter data and to make changes since values in the entire worksheet are automatically updated to reflect the changed data. Most spreadsheet packages also help users generate charts from the data and do "what if" analysis. For example, in calculating monthly mortgage payments, it takes only a few seconds to analyze what would happen if the interest rate decreased by 1 percent or increased by 2 percent.

Some tasks such as inventory management can be accomplished using a spreadsheet or a database, depending on the complexity of the data and the way it needs to be accessed and presented. Database software provides features for customizing the data entry process, sorting and searching data based on certain criteria, and generating reports. Its applications include organizing and storing address books, phone lists, recipe books, music collections, supplier lists, and employee profiles.

Today, most software is purchased on CD-ROM (a CD with a Read-Only Memory) or downloaded via the Internet. In 1981, the newly introduced IBM PC did not have a CD-ROM drive; it did not even have a hard disk drive. This meant that users had to insert the floppy disk that held the operating system every time they started the PC. They would flip on the PC's power switch and wait while the machine booted (started) and loaded the operating system (usually PC-DOS) into memory. When they wanted to run a software application, they removed the Disk Operating System (DOS) disk and inserted a program disk, such as a word processor or game.

The 1981 PC had no mouse or icons or even menus for managing software tasks. Users had to type commands, such as

Computer Measurements

bps *bits per second*
A measurement of the number of bits transferred per second on a modem.

bit *binary digit*
A bit represents a 0 or 1.

byte
A byte usually is a group of eight bits.

K or KB *kilobyte*
A kilobyte is 1,024 bytes.

M or MB *megabyte*
A megabyte is about 1 million bytes.

G or GB *gigabyte*
A gigabyte (gig-a-bite) is about 1 billion bytes.

T or TB *terabyte*
A terabyte is about one trillion bytes.

Hz *hertz*
A hertz is a unit of frequency of a single clock cycle per second. It measures how fast a computer processes data.

KHz *kilohertz*
A kilohertz is a unit of frequency of 1,000 clock cycles per second.

GHz *gigahertz*
A gigahertz is a unit of frequency of 1 billion clock cycles per second.

MHz *megahertz*
A megahertz is a unit of frequency of 1 trillion clock cycles per second.

"RUN A:WORDSTAR.EXE" to start a program. There were no colors, graphics, or sounds except for beeps from a tiny speaker in the system unit. The floppy disk could only hold 160 kilobytes (about 160,000 characters, or 80 pages) of data, and the memory held only 64 kilobytes.

In 2004, a typical PC costs about $600, less than half the cost of a 1981 IBM PC, and has a 60-gigabyte (60 billion bytes) capacity hard drive and 256 megabytes (256 million bytes) of memory. This system also includes a 15-inch color graphics monitor, CD drive, stereo sound, and a powerful multitasking processor. One computer expert illustrated the profound advances in computers since the days of mainframes by estimating that if the automobile industry had developed like the computer business, "a Rolls-Royce would now cost $2.75 and run 3 million miles on a gallon of gas."[9]

The greatest advancement, however, is in what today's PC enables the user to do. An astonishing range of creative and empowering software makes almost anything possible. The PC is a tool for business, but it is also a voice synthesizer, a stereo, a reference library, an art studio, a game center, and a chatroom.

2 The Making of the Microcomputer

AT ISSUE

As seen in the previous chapter, the personal computer has changed our lives in many ways. It was not a change planned by corporate executives, university professors, or government officials—electronic hobbyists, young people tinkering in their garages, hackers, and college dropouts initiated this transformation. Often dubbed "nerds," these individuals created the personal computer industry, and in the process, some of them became millionaires.

Bill Gates, who cofounded Microsoft at age 19, eventually became the richest person in the world. But he claims that the PC revolution was not about money. In a 2001 speech, Gates said, "The PC is clearly the most important tool to leverage mankind's creativity that we've ever had. And from the beginning, it was about empowerment, it was about the individual."[10]

How did a bunch of nerds manage to bring computing power to the individual?

BEFORE THE PC

The first mechanical counting device, the abacus, was in use over 5,000 years ago. More recent inventions include Pascal's counting wheel in 1642 and Babbage's steam-powered counting machine in the mid-1800s. Businesses used these machines to calculate freight costs, insurance rates, tide timetables, astronomical charts, and other information. As the twentieth century approached, a major advancement in speed and functionality was made.

Herman Hollerith's counting machine, a predecessor of the personal computer, calculated the data from the 1890 U.S. census in two and a half years. Ten years earlier, it had taken people eight years to process the 1880 census data by hand.

It had taken eight years for the data from the 1880 U.S. census to be tabulated. In 1890, Herman Hollerith's machines calculated the census data in just two and a half years. These new machines used a punch card system: a hole punched in a certain place on a card represented some fact about an individual. Businesses quickly adopted punch-card systems and Hollerith founded the Tabulating Machine Company, which merged with other companies and eventually formed International Business Machines (IBM) in 1924. However, the first computer did not come from IBM; it was invented by Dr. John V. Atanasoff and his graduate student Clifford E. Berry.

Atanasoff and Berry created the ABC (Atanasoff Berry Computer) to reduce the amount of time it took physics students to complete long, complicated calculations. A prototype of the ABC was assembled in 1939, and three years later a working model was finished. It used the binary numbering system, memory and logic circuits, and vacuum tubes (light bulb-size glass tubes containing filaments and circuitry). Ironically, IBM's response to the ABC was, "IBM will never be interested in an electronic computing machine."[11]

Despite those words, IBM gave $500,000 to Harvard University professor Howard Aiken to develop a computer. The Mark I, completed in 1944, stood 51 feet long and 8 feet high. IBM still was not convinced about the usefulness of computers, but the U.S. Army was interested in the technology.

During World War II, bombs often missed their targets because of bad trajectory predictions. It was difficult and time-consuming to calculate trajectories for all the weapons under various situations. The army needed a machine to make these computations, so Dr. John Mauchly and J. Presper Eckert Jr. of the University of Pennsylvania created the ENIAC (Electronic Numerical Integrator and Computer) in 1946. It was the first fully electronic computer and was a thousand times faster than its predecessors.

The ENIAC weighed 30 tons and filled 15,000 square feet of floor space. It contained 18,000 vacuum tubes that were finicky and often overheated, failing at an average rate of 1 every 7 minutes. Before it was even completed, the ENIAC was put to use making calculations for testing the atomic bomb.

After World War II, computers became big business. The first commercially available electronic computer was the UNIVAC (Universal Automatic Computer). The University of Pennsylvania's Mauchly and Eckert developed it for the Remington-Rand Corporation, which became Sperry Univac. The computer won fame during the 1952 presidential election. With only 5 percent of the votes counted, the UNIVAC predicted a sweeping victory

Harvard University's Mark I computer, completed in 1944, was 51 feet long and 8 feet tall. Although IBM had said just a few years earlier that it would never be interested in computers, IBM gave Harvard professor Howard Aiken $500,000 to develop the Mark I.

for Dwight Eisenhower over Adlai Stevenson. News anchor Walter Cronkite hesitantly announced the prediction to the world. The UNIVAC proved to be correct and the world was changed; suddenly, computers were chic.

To most Americans, the word "computer" meant the UNIVAC because it was the only machine they knew. Before long, IBM realized that they should develop large mainframe computers, too. They introduced the IBM 701 in 1953 and the IBM 650 in 1954. Sales estimates for the IBM 650 were optimistically set at 50 units; about 1,000 were installed. The IBM 650 became the machine that set the standard— "the Model T of the computer industry"—and IBM took a dominant position in the industry.

The Electronic Numerical Integrator and Computer, or ENIAC, was the first fully electronic computer. Built at the University of Pennsylvania in 1946 for the U.S. Army, it weighed 30 tons and contained 18,000 vacuum tubes.

With big mainframe computers, everyone had to share computer time. Those computers were too expensive for one person to own. Programmers had to make appointments and then submit their jobs on a stack of punch cards to an operator. If the program had problems, it had to be fixed and then another appointment made to run the revised program. By the late 1960s, several users could submit jobs simultaneously by working at terminals connected to the machine. Time-sharing came into vogue, allowing smaller organizations to have their own terminals and buy time on large machines. Still, computer access was limited and required special programming skills.

Meanwhile, excitement was growing about a tiny electrical device called a transistor, which had been invented in 1947. A transistor is composed of semiconductor material (silicon) that can both conduct and insulate. It is smaller than the vacuum tube and does not generate as much heat or burn out as rapidly. Devices that integrate a number of transistors into a more complex circuit are called integrated circuits or chips. In the 1960s, most of the companies making silicon chips were in California's Santa Clara Valley, and the area came to be known as Silicon Valley.

Silicon chips made computers more powerful, more reliable, less expensive, and cooler to operate. Using silicon technology, Digital Equipment Corporation built the first minicomputer. When introduced in 1963, the PDP-8 minicomputer cost only $18,000, compared to $100,000 or more for a mainframe. Rather than filling an entire room, it fit in the corner of an office and in the backseat of a Volkswagen Beetle convertible, as shown in advertisements.

The little PDP-8 sold in big quantities. It was installed in laboratories, newspaper offices, schools, and even ran the digital scoreboard at Fenway Park in Boston. Its success signaled the increasing presence of a demand for smaller computers.

THE ERA OF ALTAIR AND APPLE

As the components grew smaller and less expensive, it seemed possible that a computer could be made small enough to fit on top of a desk. The missing key element was a small, general-purpose processor. The solution to that problem came from the newly founded Intel Development Corporation of Silicon Valley.

In the summer of 1969, a Japanese company asked Intel to produce a set of chips for a new line of programmable electronic calculators. Intel engineer Marcian E. Hoff studied the plans for the calculator and concluded that the design was too complex. He proposed that Intel develop a general-purpose logic chip, "a processor chip," rather than chips specific to only one task.

Like the central processor of a mainframe, the microprocessor would be programmable. This meant that Intel would not have to create a custom set of chips for every customer. If the processor needed to be a calculator, a program would make it a calculator. If it needed to be a digital clock, a program would make it a clock. It was a brilliant idea.

The first microprocessor, named the 4004, was developed in late 1970. It had 2,250 transistors, processed four bits of data at a time, and performed about 60,000 operations per second. By today's standard, it was a clunker, but for its era, it was quite extraordinary.

At first the Japanese company prevented Intel from selling the chip to other companies. But in 1971, Intel gained the rights to sell it on the open market. With no idea of the possible market for the chip, Intel introduced the 4004 and its associated RAM, ROM, and input/output chips. The four chips were mounted on a small circuit board—about the size of a pocketbook—called "the motherboard" because it was the main board.

In April 1972, Intel introduced the 8008 chip, the first "eight-bit" microprocessor, which means that it could process eight bits of data at a time. Two years later, Intel released the powerful 8080. Yet, Intel management did not recognize the potential of what it had created. Gordon Moore, cofounder of Intel, recalled that an engineer presented an idea for a home computer: "And while he felt very strongly about it, the only example of what it was good for that he could come up with was the housewife could keep her recipes on it," says Moore. "And I couldn't imagine my wife with her recipes on a computer in the kitchen. It just didn't seem like it had any practical application at all, so Intel didn't pursue that idea." [12]

Other companies, such as IBM and Digital, could have made a personal computer, too. Engineers knew it was technically feasible, but executives at these companies did not foresee a market for such a device. Ken Olsen, founder of Digital, asked, "Why would anyone want a computer on his desktop?" [13] It was not corporations or universities that created the personal computer, it was electronics hobbyists.

Intel developed the 4004 microprocessor, at left, in 1970. Offered for sale on the open market in 1971, the first microprocessor had 2,250 transistors, processed four bits of data at a time, and performed about 60,000 operations per second. At right is Intel's second microprocessor, the 8008 chip, which processed eight bits of data at a time.

As soon as Intel made the 4004 and 8008 chips available, technically savvy hobbyists fulfilled their dream of building their own computers. A few decided to package the pieces in a kit and sell them to less technical people who also wanted a computer. News of one of the first computer kits was presented on the front page of the July 1974 issue of *Radio Electronics.* The headline read: "Build the Mark-8, Your Personal Minicomputer." (The term "personal computer" did not exist yet.)

The Mark-8 machine consisted of six circuit boards, one of which held Intel's 8008 processor. The user had to assemble the boards into a case with switches and lights across the front of it. To make the machine do calculations or play games, the user entered instructions one bit at a time by flipping switches.

Results were displayed on the panel of lights next to the switches. All instructions and data were lost every time the machine was powered off.

Despite the tedious interface and limited functionality, about 2,500 people bought the Mark-8 boards. A number of user groups formed to share hints and tips about building and using the machine.

Six months later, in January 1975, the cover of *Popular Electronics* featured the Altair 8800. The Altair was based on the more powerful 8080 processor and designed by a small electronics firm called Micro Instrumentation and Telemetry Systems (MITS) in Albuquerque, New Mexico. This computer kit sold for $395; fully assembled, the Altair cost $650.

Bringing the Altair to market had not been easy for MITS founder Edward Roberts. As development neared completion, Roberts was $300,000 in debt and about to go bankrupt. To his amazement, he received a $65,000 loan from a bank on the flimsy promise of sales from this new computer kit. He took the money and finished developing the computer, but then could not decide what to call it.

Les Solomon, the editor of *Popular Electronics,* was writing a feature article about the kit and needed to know the name of it for the magazine cover. Roberts told Solomon, "I don't give a damn what you call it. If we don't sell 200, we're finished." [14]

Solomon searched for a name. One day he noticed his 12-year-old daughter watching *Star Trek* on television. He asked her what the name of the computer on the *Enterprise* was and she told him it was named Computer—a name that clearly wouldn't work. His daughter asked, "Why don't you call it Altair? That's where the *Enterprise* is going tonight." [15]

The Altair sold like crazy. Within three months, 4,000 people placed orders. MITS could not build kits, peripherals, or memory boards fast enough. The tiny company was still operating from a few rooms in a building next to a laundromat in an Albuquerque shopping center.

In addition to the demand for hardware, customers needed a BASIC interpreter. This would enable Altair 8800 users to write programs in the high-level language of BASIC rather than using the peculiar and crude commands of machine code. Two nerds stepped in and provided the solution.

In a period of just five weeks, Bill Gates and Paul Allen developed a version of BASIC for the Altair. They did not invent BASIC; it had been developed in 1964 by Dr. Thomas Kurtz and Dr. John Kennedy of Dartmouth College. Gates and Allen did what they were good at—they saw an opportunity and pulled together the resources to execute their idea.

Allen quit his job to become the software director for MITS and Gates dropped out of Harvard to write software for MITS and other companies. The two young men later formed Microsoft Corporation and moved back to their hometown, a suburb of Seattle, Washington. Microsoft became the largest and most influential software company in the world and Gates became the richest man in the world.

Of his abandoning college for the business world, Gates said "I never really made a conscious decision to forgo a degree. Technically, I'm just on a really long leave. Unlike some students, I loved college. . . . However, I felt the window of opportunity to start a software company might not open again. So I dove into the world of business when I was 19 years old."[16]

As Microsoft jumped into the software business, lots of other companies dove into the hardware business. These included IMSAI, Commodore, Radio Shack, North Star, Apple, Itty Bitty Machine Company, and Kentucky Fried Computers. Few survived, including MITS. Roberts sold MITS for $6 million, bought a farm, and went to medical school. Under new management, MITS collapsed.

One of the success stories is Apple Computer. It all began when Steve Wozniak and Steve Jobs showed their prototype Apple I at a meeting of the Homebrew Computer Club. The machine was mounted on a sheet of plywood with all the

The popular Apple II, introduced by Apple Computer in 1977, was the first personal computer to have a floppy disk drive, allowing data to be shared with other Apple IIs, and a color monitor. By 1980, more than 120,000 of the PCs were sold.

components visible, and when a local computer dealer saw it, he ordered 100 units.

It seemed the Apple was a winner, if the young men could get the money to actually produce the computers. Jobs sold his Volkswagen bus, Wozniak sold his Hewlett-Packard programmable calculator, and they borrowed $5,000 from a friend. On April Fools' Day, 1976, Jobs and Wozniak founded the Apple Computer company and released the Apple I. This computer did not use the Intel microprocessor; it used the 6502 processor chip, from MOS Technology. It had no keyboard, case, or sound, and sold for the strange price of $666.66.

Like most PCs at the time, the Apple I used a tape recorder for data storage. However, the very next year, the Apple II was introduced and it had a floppy disk. It also had a color monitor, which appealed to game players, and ran the spreadsheet program VisiCalc, which appealed to business users. By the end of 1980, over 120,000 Apple IIs were sold.

Apple sales rose from $775,000 in 1977 to $7 million in 1978 to $335 million just three years later. It became the fastest-growing company in American history. Suddenly, the nerd-filled industry of personal computers grabbed the attention of corporate America.

THE BIRTH AND CLONING OF THE IBM PC

In July of 1980, a group of 12 IBM engineers secretly assembled in Boca Raton, Florida, to design and build a new personal computer code-named "Acorn." They wanted to bring this computer to market in just one year. Thus, IBM had to abandon its traditional policy of creating its own proprietary hardware and software.

IBM built the computer from off-the-shelf components. They bought microprocessors from Intel—the new 8088, a 16-bit chip. They added 16 kilobytes (K) of memory, floppy disk drives, and a monitor, and called it the IBM PC.

All that hardware was useless without software, so IBM contacted Microsoft to create an operating system. Since Gates had never written an operating system, he suggested they talk to Gary Kildall of Digital Research, who had written the operating system CP/M (Control Program for Microcomputers). However, things did not work out with Kildall. If they had, the software industry might look significantly different today; perhaps CP/M would be running on the majority of PCs.

IBM went back to Microsoft and gave them the contract to develop the operating system, along with a very tight schedule. Gates immediately hired Tim Paterson, who had written QDOS, the so-called Quick and Dirty Operating System. Paterson adapted QDOS to work on the IBM computer. The finished product was called MS-DOS (Microsoft's Disk Operating System) and the version for the IBM PC was called PC-DOS.

The schedule was met, and on August 12, 1981, IBM unveiled the IBM PC. *Newsweek* reported, "For years futurists have been predicting the advent of a world in which computers are as common as can openers. . . . IBM is putting its stamp of approval on the microcomputer and that means that it's here to stay." [17]

In the first year, about 35,000 units were sold at a base price of $1,365. By the end of 1982, over 800,000 had been sold. There was no technological marvel to account for the success of the IBM PC. Its success was largely based on the fact that it was a product of IBM, which had more than 80 percent of the market share of large computers and was one of the biggest and wealthiest companies in the world.

The success of the IBM PC brought success to many other companies. Since IBM marketed the PC through outside distributors such as Sears and ComputerLand, the PC retail industry experienced rapid growth. So did the market for peripherals like printers, joysticks, and modems. Since anyone could learn the IBM PC's design, anyone could build accessories for the machine and go into business.

As for software, PC-DOS became the standard operating system for the IBM PC. Software developers jumped at the opportunity to write applications to run with DOS. For example, Mitchell Kapor founded Lotus Development Corporation and introduced Lotus 1-2-3. This electronic spreadsheet made the IBM PC a valuable machine in the business marketplace.

Small computer companies sprang up and tried to sell machines similar to the IBM-PC. It was relatively easy to build a clone because the microprocessor and other chips

Early on, IBM cornered the PC market on the strength of the company's success with large computers and reputation as one of the biggest, wealthiest companies in the world. The IBM PC was introduced in August 1981, and by the end of the following year, more than 800,000 of the machines had been sold.

were for sale. But customers wanted the clones to work exactly like the IBM PC and run the same software applications. To accomplish this, programmers had to figure out some low-level code (called BIOS code) in the IBM PC. Compaq Computer was the first to build a machine that was 100 percent compatible. It was also a portable computer, weighing only 28 pounds. In its first year, Compaq sales totaled more than $100 million.

Many other compatible brands followed. Tandy, Zenith, Sperry, Osborne, and Leading Edge offered PCs. They were priced lower than the IBM PC, but ran the same applications—

and that was what was important to customers. Slowly, the compatible machines whittled away at IBM's control of the personal computer market.

IBM kept one step ahead of the clones by releasing a new model of PC with the newest technology before the other companies. In the spring of 1983, IBM released its PC/XT. It had a hard disk that could store 10 megabytes of data, about 10 million characters.

Eventually though, IBM slipped. They held back on releasing a PC model with a new 386-bit processor in it, fearing that it was so powerful it would compete with their own minicomputers. Meanwhile, Compaq went ahead and, in 1986, became the first company to announce a 386-based computer. This gave Compaq an image of leadership that once had belonged solely to IBM.

IBM tried to eliminate the competition by redesigning the PC hardware and replacing MS-DOS with OS/2. The new machine, called the PS/2, could not be cloned. But, unfortunately for IBM, many customers did not like the PS/2. The operating system was complicated and cumbersome, and accessory cards for the PS/2 were few. The PS/2 design did not become the hardware standard.

The clones continued to take advantage of IBM's missteps. A company called Dell Computer was founded in 1984 in a college dorm room. Michael Dell offered inexpensive compatible PCs and within five years had sales of $250 million. By 1990, Dell and Compaq were earning more profits from the PC market than IBM did.

The success of the PC market meant success for chip manufacturers, too, especially for the makers of microprocessors. A few companies began cloning Intel's processor chips, but Intel remained the industry leader by continuing to produce faster and more powerful chips one after the other. By the end of 1983, Intel had 21,500 employees and $1.1 billion in annual sales.

Digital Devices

Although computers seem like complex devices, their underlying principle is quite simple. It is based on electrical circuits being open or closed (on or off). One state represents the digit 1 and the other state represents the digit 0. Thus, computers are based on the binary (base 2) number system, which consists of two digits: 0 and 1.

A 0 or 1 is called a bit, for binary digit. A group of eight bits is called a byte and typically is used to represent a single character. For example, in the popular coding scheme ASCII, the letter *M* is represented as 01001101. When the user presses *M* on the keyboard, the code 01001101 is sent to the computer's memory. The internal workings of a PC know nothing about *M*s or *P*s or *Z*s, they only know about 0s and 1s.

The word "digital" comes from the word "digit," referring to human fingers. Today, digital refers to devices that represent signals or information using a binary method. Digital data consist of data that are in discrete, discontinuous form.

Analog data are in a form that has variable values and represents the real world. Most natural phenomena, such as temperature, sound, light, and air pressure, are analog. Telephone, television, radio, and cable TV historically have been transmitted as analog signals, but now many are switching to digital.

Since old phone lines used analog signals, PCs traditionally needed a modem in order to access the Internet via a phone connection. A modem (modulator-demodulator) had to convert the digital data of the PC to analog data. The data were then sent over the phone line and at the receiving end the signal was converted from analog back to digital.

Today, many analog realities are connected to digital representations. For example, music can be sampled and the sounds converted to numbers. The numbers are averaged and then compressed. The digital process produces a copy that sounds nearly identical to the original music.

GETTING GUI

Instead of trying to copy IBM, Apple Computer took a different approach to attracting customers. Its machines did not use Intel microprocessors; they used processors from MOS Technology and Motorola. In addition, Apple computers did not run MS-DOS; they used Apple's own operating system and a graphical user interface, or GUI (pronounced GOO-ee).

The first GUIs came from work at the Xerox Palo Alto Research Center (PARC). Engineers at PARC had experimented with a number of different mouse configurations as an addition to the keyboard. The screen contained menus and little pictures called icons, which a user pointed at and then selected by clicking the mouse. It made the computer easy and friendly because users did not have to learn and remember the commands—such as CTRL+P to print and ALT+SHIFT+F2 to save—that were used with PC-DOS.

In January 1983, Apple introduced the Lisa home computer, named after the daughter of one of the designers. It was the first PC to provide a GUI. Users liked clicking on and dragging things with the mouse. They were also impressed with the pull-down menus and many other friendly features, but the price was a whopping $9,995. Only 10,000 units were sold before the Lisa was scrapped.

Later, in 1984, Apple wowed the public with the cute, little Macintosh. It was inexpensive and fun to use. The Macintosh GUI was embedded in the operating system, which meant all applications had a similar user interface. For example, users did not have to learn different printing commands for their word processor, spreadsheet, and database programs; there was one print command that worked with all applications.

The Macintosh shifted the perception about personal computers. Although the circuitry on the inside worked pretty much the same as the IBM PC, it looked and felt different on the outside. Nonhackers, nonscientists, and even children could use this machine. Whereas the IBM PC was seen as a complex machine suited mostly for business people, the Macintosh appealed to students, home office users, and computer graphic designers.

Several companies, including Microsoft, tried to design GUIs for MS-DOS machines. Microsoft Windows version 1.0 was introduced in 1985 but not widely accepted. Users who knew the DOS command language could work more efficiently with the command interface and resisted changing to Windows. Also, Windows required considerable processor power and a high-quality monitor. These were still very expensive.

Microsoft had other problems, too. Apple sued Microsoft, claiming that Windows 1.0 had the "look and feel" of a Macintosh. The lawsuit ended when Microsoft agreed not to use Macintosh technology in Windows version 1.0. However, there was no restriction placed on future versions. With Windows 3.0, Microsoft had a GUI with pull-down menus, scroll bars, clickable buttons, multiple windows, and icons. Users liked it, and hundreds of new Windows-based application programs followed.

The PC had become friendly, and some 65 million PCs were in use by 1992. Changes in technology continued to improve functionality and performance. Intel's Pentium microprocessor, released in 1993, contained 3.1 million transistors and operated twice as fast as its predecessor, the 486. It enabled a PC to better handle motion video, animation, graphics, sound, and other media.

In fewer than 20 years, the personal computer evolved from being a toy for nerds to being a serious but friendly tool in offices, schools, and homes around the world. According to a report from Gartner Dataquest, the billionth PC shipped in April 2002.

3

The PC
Means Business

AT ISSUE

In *Accidental Empires,* Robert X. Cringely pokes fun at the PC and its influence on the business world: "PCs killed the office typewriter, made most secretaries obsolete, and made it possible for a 27-year-old M.B.A. with a PC, a spreadsheet program, and three pieces of questionable data to talk his bosses into looting the company pension plan and doing a leveraged buy-out." [18]

Then Cringely admits: "After automobiles, energy production, and illegal drugs, personal computers are the largest manufacturing industry in the world and one of the great success stories for American business." [19] In fact, PCs created one of the longest continuous peacetime economic expansions in U.S. history.

RAISING AN INDUSTRY

When the first microcomputers came along, most business managers dismissed the machines as glorified typewriters or calculators on steroids. Early business applications were limited to word processing to create and edit documents, basic spreadsheets to keep accounting ledgers, and simple databases to manage inventory, clients, and other records.

As processor speed and power increased, software applications improved and transformed the business world. Today's powerful software, running on a Pentium 4 system and viewed on a large color monitor, makes creating and managing letters, brochures, financial reports, and business presentations a whole new experience. In just about every office, there is a PC on the desk.

Personal computers are not only the tools of Big Business, they are a big business in themselves, as this photo of the Dell Computer Corporation's huge manufacturing facility in Austin, Texas shows. PCs created one of the longest continuous peacetime economic expansions in U.S. history.

The PC's usefulness in business is obvious today, but it took time and action to convince business people to give the PC a test drive. In the late 1970s and early 1980s, people still considered the PC a toy. About 20 different companies were selling some 250 different computer games. Pong, an electronic version of table tennis, was one of the first popular game programs.

The computer game market had sales of about $2 million in 1982. Some estimates say that more than half of the PCs bought for home use were devoted mainly to games. These games may or may not have had any educational value, but they did bring PCs into the home and convince people that the sterile-looking machines could be fun to use. Kids were not only fascinated by the blips moving back and forth

Spreadsheet programs were instrumental in convincing business owners that the PC was essential to their operations. VisiCorp, whose founder, Dan Fylstra, is shown here, marketed the first commercial spreadsheet programs, VisiOn Calculator for the IBM PC XT (running on the PC on the left) and VisiCalc for the Apple II (at right).

across the screen, they were also intrigued by the world of computer programming.

"Games aid in the discovery process," said Philip D. Estridge, head of IBM's PC operations.[20]

After playing some games and getting comfortable with the machine, some people discovered the PC's potential for work. For example, the Brown family of Minneapolis found many uses for their home computer: the daughter stored class notes for school, the father filed names and addresses of business clients, the son wrote programs, and the mother managed the records of a gourmet cookware store. Aaron Brown, the father, said, "It's become kind of like the bathroom. If someone is using it, you wait your turn."[21]

VisiCalc, which stands for visual calculator, was the first software application to show businesses that a PC could do more than BASIC programming and games. Released in 1979, the spreadsheet program ran on the Apple II computer. *Byte* magazine wrote: "Quite possibly the program responsible for the '80's Wall Street frenzy: VisiCalc on the Apple II."[22]

Companies spent a great deal of time doing financial projections, which required lots of manual calculations. Changing a single value in an accounting or budget worksheet meant recalculating all of the values in the sheet. With VisiCalc, if one number was changed, all the numbers that depended on it would be recalculated automatically.

"VisiCalc took 20 hours of work per week for some people and turned it out in 15 minutes and let them become much more creative," says Dan Bricklin, who devised the program while attending Harvard Business School.[23]

By 1982, the spreadsheet was the "killer application"—the application that justified the purchase of a PC for office work. In addition to VisiCalc, there was now Lotus 1-2-3, which had graphics and data retrieval functions. Lotus 1-2-3 quickly became the favorite of IBM PC users. About 1 million spreadsheet programs were sold in 1984.

In his article "A Spreadsheet Way of Knowledge," Steven Levy wrote:

> There are corporate executives, wholesalers, retailers, and small-business owners who talk about their business lives in two time periods: before and after the electronic spreadsheet. They cite prodigious gains in productivity. They speak of having a better handle on their businesses, of knowing more and planning better, or approaching their work more imaginatively. A virtual cult of the spreadsheet has formed, complete with gurus and initiates, detailed lore, arcane rituals—and an unshakable belief that the way the world works can be embodied in rows and columns of numbers and formulas. [24]

Word processing became another popular application. While writing programs on the Altair computer, Michael Shrayer decided to write the manuals on the same machine. He developed Electric Pencil, the first microcomputer word processor. Soon, WordStar came along and impressed users with its extensive capabilities. In 1986, *PC Magazine* reviewed 57 different word-processing packages.

Lawyers, small business managers, writers, and doctors discovered how PCs could help them with their work. Even one pig farmer grasped the advantage of having a PC. He used it to access the farm bureau's service, which provided weather conditions, up-to-date hog prices, and general advice. He also computerized all his farm records. "This way, you can charge your hogs the cost of the feed when you sell them and figure out if you're making any money," he said. "We never had this kind of information before. It would have taken too long to calculate." [25]

The IBM PC brought power to people. It did not matter to these people that this machine was not some marvel of technological distinction. It did not matter that it was built from cheap, off-the-shelf parts. According to IBM Fellow Mark

Dean, who designed the color-graphics adapter for the original PC, "It wasn't rocket science, and it wasn't extraordinarily inventive. Most of the time it's not the level of technology that makes a difference. The important thing is that you are solving a problem at the right time and at the right price."[26]

Over the years, engineers have criticized PC hardware and software. Sun Microsystems CEO Scott McNealy even called the Microsoft Office suite of applications "a hairball." But despite their many flaws, the basic designs of these machines and their software have worked pretty well. Stephen Wildstrom tries to explain this odd situation: "The computers are still too hard to use. Software quality leaves much to be desired, and consumers are expected to put up with a level of failure and frustration that they would not accept from any other product. Yet I have to admit that the reliability of the hardware and the usability of the software have improved dramatically over two decades."[27]

The first PCs were all slightly different and so the software for each was slightly different. Programs written for one machine could not run on another brand of machine, keeping customers faithful to a hardware firm. There was a push toward "open" standards for both software and hardware, as the IBM PC's hardware was. The IBM operating system was not completely open, but it was open enough for non-Microsoft application programmers to create products.

These open standards meant that as the IBM PC succeeded, new markets and opportunities opened for other players. Software developers, computer retailers, chip manufacturers, and hardware makers could all take advantage of the rapidly growing market of microcomputer buyers. The computer industry was irrevocably changed.

Going into the software business was easy. It cost almost nothing. Writing software took only time and skill. Distribution cost very little or nothing. A program could be copied onto a disk and then distributed through user groups, computer stores, or electronics shops. A cheaper way to distribute software was through online bulletin boards; users who wanted the program simply downloaded a copy. Word of mouth sold the program, not expensive or elaborate advertising.

Some programmers gave away their programs for free and called it "freeware." Other groups of programmers let users try it for free, but asked them to send a payment if they liked it. This was called "shareware." The idea of charging money for software and selling it in retail stores was a new and foreign idea to most nerds, but not to enterprising businesspeople.

In 1983, Ed Faber thought the time for retail computer stores had arrived, so he signed on as the first CEO of ComputerLand. "My most ambitious hopes were that we'd have some store somewhere that might be able to do $50,000 a month," he said. "Well, the average ComputerLand store does $130,000 a month." [28]

The PC industry spurred publishers to create several new magazines geared for computer users. To help its readers evaluate PCs in the same way other magazines helped readers evaluate car models, *PC Magazine* began doing performance and compatibility tests on hardware and software. These independent tests were done in the *PC Magazine* labs, known as the Toy Shop.

One unforgettable test was performed on a surge suppressor, which is used to protect a PC from surges of electricity coming over the power line. The test engineer took off his rings and watch and stood on a rubber mat. "[He] loaded the surge suppressor into a big black case, lowered a Lexan shield over the case, and zapped it with an artificial lightning bolt from the surge generator," reported *PC Magazine*. His notes on the test read: "Device emitted orange flame and gray smoke. Device did not work subsequent to test." [29]

MONOPOLY

"The founders of the microcomputer industry were groups of boys who banded together to give themselves power," Cringely wrote in *Accidental Empires*. "They defined, built, and controlled (and still control) an entire universe in a box. . . . And turning this culture into a business? That was just a happy accident that allowed these boys to put off forever the horror age—that dividing line to adulthood that they would otherwise have been forced to cross after college."[30]

These boys did not have to grow up or conform to any code of conduct. Almost overnight, they went from being social outcasts to having the world lay money and fame at their feet. Most of them were still rebellious juveniles who had no idea how businesses were supposed to work, yet they were running million-dollar software companies. It was "the triumph of the nerds."[31]

The nerds had no sense of what was possible or impossible, so they did business in whatever way suited them. They wrote software and gave it away or trusted the people who liked it to send them money. Then, they worked 15 hours or more a day writing more software—coding the next release or the next ingenious product as soon as one went out the door, trying to stay one step ahead of the hotshot programmer down the street.

Winners rose to the top and then flopped six months later, when new, faster, more feature-packed products hit the market. The first big-winning application, VisiCalc, was beat out by Lotus 1-2-3, which was later too slow to release a Windows version and thus gave Microsoft Excel the opportunity to become the most popular spreadsheet. WordStar became more popular than Electric Pencil, but was later replaced by WordPerfect, which then lost the consumers' favor to Word. Then, Microsoft bundled Word, Excel, Access, and PowerPoint into Microsoft Office. The integrated suite of programs became the most widely used application on new PCs.

"In the PC business, constant change is the only norm, and adolescent energy is the source of that change," says Cringely.[32] The corporate culture at PC businesses—volleyball, junk food, T-shirts, and hundred-hour workweeks—was very different from that at IBM, General Motors, or Japanese electronics firms. For example, after shipping a new product, the employees of

Adobe Systems Inc., a software company, celebrated with a companywide water fight.

Going to work for Microsoft was almost like joining a cult. Gates did not hire experienced and skilled programmers. Instead, he hired computer science graduates right out of college, relocated them to the Seattle area, and indoctrinated them into the Microsoft way of working and thinking. Cringely describes the experience of these programmers:

> So here are these thousands of neophyte programmers, away from home in their first working situation. All their friends are Microsoft programmers. Bill is a father/folk hero. All they talk about is what Bill said yesterday and what Bill did last week. And since they don't have much to do except talk about Bill and work, there you find them at 2:00 A.M., writing code between hockey matches in the hallway. [33]

Microsoft employees developed an unflagging dedication to the company, and this gave the company a tremendous advantage. Microsoft accomplished things that seemed impossible in the 1970s. In essence, they were the key company that led the PC industry to play such an important role in the U.S. and world economy.

In the process, Microsoft became a very large and very influential company—in fact, it held a monopoly on PC operating systems. In the late 1970s, Gates often said, "We want to monopolize the software business." [34] By the 1980s, however, he realized that he had to stop saying it because competitors were crying "foul" and antitrust lawyers had started eyeing the company.

In 1998, the U.S. Department of Justice and attorneys from several states filed suit against Microsoft. They claimed that the inclusion of the Internet Explorer World Wide Web browser in the Windows 98 operating system violated antitrust guidelines, and in April 2000, a federal district judge found Microsoft guilty

The 1998 lawsuit brought against Microsoft by the U.S. Department of Justice and representatives from the offices of several state attorney generals contended that the company had violated antitrust laws. Microsoft's upper management, including Bill Gates, seen here at the podium, fought hard to keep the courts from splitting the company into two independent parts.

of these violations. Two months later, he ordered the company to be split into two parts.

The announcement sent shock waves through the entire industry. Would this economic engine be torn apart? It did not make sense to many economists. And what was the significance of any monopoly in the ever-changing world of computers? IBM had had a monopoly in the mainframe industry at one time, yet that advantage did not account for much in a matter of a dozen years. However, others felt Microsoft was using its position to bully distributors and intimidate its software competitors.

Implementation of the order against Microsoft was stayed while Microsoft appealed the decision. In June 2001, a federal appeals court issued an opinion skeptical of the breakup remedy, and in September the administration of President George W. Bush decided not to pursue a breakup.

The Department of Justice's antitrust suit against Microsoft finally ended in November 2002. The result was that Microsoft had to disclose the internal software codes of some products, stop some special pricing, allow non-Microsoft applications to launch automatically, and make a few other changes. However, the company would not be split into independent pieces and it did not have to break apart the Windows operating system. Microsoft was allowed to maintain its monopoly on desktop PC operating systems.

The annals of Microsoft clearly testify to the new rules of competition and the new way of doing business. Two nerds stepped out of nowhere, outmaneuvered and outpaced their competitors, and created a software empire with annual sales of $11 billion. The stock market values the company at well over $150 billion—far more than either IBM or General Motors. Yet Microsoft's employees still walk around in T-shirts.

LIFE ACCORDING TO MOORE'S LAW

In 1989, the PC industry grew at a pace in accordance with Moore's Law, which became the yardstick by which the PC industry is measured. However, the law originated long before the PC did, and was not conceived through scientific analysis. The author of Moore's Law, Intel cofounder Gordon Moore, explained to *PC Magazine*:

> I was writing an article for the 35th anniversary edition of another magazine—*Electronics*—and they wanted me to predict the future of semiconductor components for the next ten years. . . . I call that Year Zero, 1959 [the year with one planar transistor]. We'd gotten up to 64 in six years—in 1965. So I said, "Aha, it's been doubling every year." I just said, "Okay, it's going to continue to do that for ten years." [35]

Over the next 10 years, the forecast held up. At that time, Moore predicted that the capacity of computer chips would

Intel cofounder Gordon Moore deduced what became known as Moore's Law, which says that the number of transistors in a microprocessor will double every 18 months. Moore, seen here, retired in 2001 after reaching the company board of directors' mandatory retirement age of 72.

double every two years. Again, his prediction proved true. The average of his two predictions—a doubling every 18 months—was what came to be known as Moore's Law. In accordance with the law, Intel microprocessors have doubled in their transistor count every 18 months. Variations of the law held that the same would be true for memory chip capacity and processing power.

These perpetual technological advances were instrumental to the growth of the PC industry. Every year, PC hardware makers had to convince consumers that their old model was outdated and they needed a new model with more power and more memory. And every year, the software companies came out

with new versions of their application packages. The new software had features that users could not live without, even though
they had been happy with the previous version. This business
strategy ensured a continuing stream of sales and profits.

Worldwide shipments of PCs climbed from 4.8 million in
1982 to 21 million in 1989. Nine million of those machines were
purchased in the United States. "The personal computer industry
had become big business, with ceaseless litigation and the
focused attention of Wall Street; and this technology, pioneered

Intel Microprocessors

MODEL #	YEAR RELEASED	# OF TRANSISTORS
4004	1971	2,250
8008	1972	2,500
8080	1974	5,000
8086	1978	29,000
80286 (or 286)	1982	120,000
80386 (or 386)	1985	275,000
80486 (or 486)	1989	1,180,000
Pentium	1993	3,100,000
Pentium II	1997	7,500,000
Pentium III	1999	24,000,000
Pentium 4	2000	42,000,000

in garages and on kitchen tables, was driving the strongest, most sustained economic growth in memory."[36]

Companies were downsizing their computer systems. Mainframes, supercomputers, and midrange computers had owned two-thirds of the computer market in 1987, but revenue from the PC market surpassed the combined market for mainframes and minicomputers in 1991. Personal computers were no longer a niche in the computer industry; they were becoming its main product.

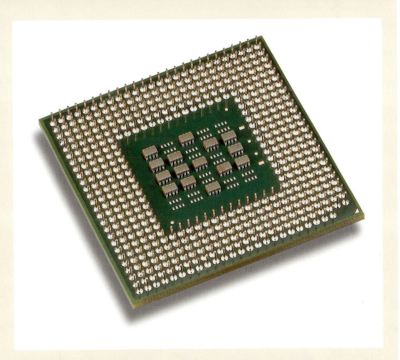

Intel's Pentium 4 microprocessor chip, first released in 2000, has 42,000,000 transistors. Today's graphics-heavy software wouldn't be able to run without powerful microprocessors like the Pentium 4.

IBM, Digital Equipment, Data General, Unisys, and other companies that made large computers were in a state of upheaval. PC makers did better as networks of PCs replaced mainframes. However, even though sales of PCs were up, profits were not. Price wars had erupted in the industry and, during 1992, the cost of a typical PC was halved.

Intel, one of the companies that manufactured microprocessor chips, fared quite well. The 486DX chip cost around $20 to make and sold for $300. Even though competition had forced lower prices, Intel still had a net profit margin of 17 percent. They had 80 percent of the market for microprocessors used in IBM-compatible PCs. Part of their success came from clever advertising that educated nontechnical consumers about the significance of the brand of microprocessor in their computers. They convinced people to buy a PC with an "Intel inside."

In 1998, Apple showed the world a "different" PC when it introduced the iMac. It had a color case and a sense of style that PC customers had never seen. They loved it, and for several months, the iMac became the best-selling computer on the market.

There was double-digit growth in the PC market in the late 1990s. This was mainly caused by the increased demand for Internet access and the Y2K (year 2000) spending binge. Many older computers would not be able to handle the date change from 1999 to 2000, so they were replaced in order to prevent problems. During this buying binge, Dell became the leading seller of PCs with a 19.8 percent market share. Compaq was second with 15.9 percent, followed by Hewlett-Packard at 11.6 percent, Gateway at 8.9 percent, IBM at 5.5 percent, and Apple at 3.9 percent.

Since then, the U.S. PC market has slumped, experiencing negative growth for the first time in history. The PCs bought in the late 1990s are due to be replaced, but "the sticky wicket is that users may be hanging on to them longer than they normally would because they are worried about the economy."[37] Also, no

Apple Computer's iMac showed the company's creative bent with its funky design and choice of five colors for the case, seen here from above. Consumers loved it, and Apple has continued to develop innovative designs for other models.

new technological advancements have been introduced to the market that have been so marvelous as to convince consumers to overcome their money worries.

Research analyst David Bailey says, "Most of the people who want PCs have PCs. It may be difficult in this economic environment to sell what remains largely a big ticket, discretionary item." [38] Even if sales do pick up, profits may not increase. As of August 2, 2002, the average selling price for desktop units was $801, the lowest in months.

"The public's infatuation with all things digital has faded," concludes John Heilemann. "On Wall Street and on Main Street, in the press and even in Silicon Valley, a feeling has set in that the information revolution has played itself out, or at least has entered a period of prolonged abeyance [temporary inactivity]."[39]

In a recent article in *The New York Times,* David Brooks wrote, "Suddenly, it doesn't really matter much if the speed of microprocessors doubles with the square root of every lunar eclipse (or whatever Moore's Law was). . . . Of course, people are still using computers. . . . What's gone is the sense that the people using the stuff are on the cutting edge of history, and everyone else is road kill."[40]

THE NEW WAY OF WORKING

Typewriters were replaced by dedicated electronic word processors in the late 1970s to early 1980s. Secretaries resisted being placed in typing pools, where the work was less interesting. Managers resisted losing control over typing work and losing a status symbol—a secretary. In the mid-1980s, dedicated word processors were replaced by personal computers and, in many cases, secretaries were replaced by PCs. The manager was given a PC and expected to do his own typing.

It was predicted that by 1990, the traditional secretary would be extinct. It was also predicted that PCs would increase productivity among workers. Is there any hard evidence that PCs increase office productivity the way executives hoped they would?

Workers spend hours every year tinkering with fonts and graphics, making multiple drafts of memos, and playing games. They spend days recovering from system crashes. They spend weeks learning complex software operations, sometimes more time than the programs are worth in savings. And those programs become obsolete in years, or even a few months.

The total cost of ownership of a typical PC is more than just hardware and software. There is also training, support, maintenance, and other expenses, which can add up to a total of $6,000 to $13,000 per PC each year.

Perhaps more work is getting done, but not in the office. Home PCs, laptops, and handheld PCs allow workers to leave their desks and take their work with them. According to the industry research company IDC, the mobile and remote workforce in the United States was expected to reach 50 million by the end of 2002. The strong division between work and home is gone. Today's technology makes it possible to work at home, on airplanes, in trains—just about anywhere. The problem this creates is that some people work everywhere, all the time. There is no downtime; there is burnout.

For some people, though, PCs and remote access are a completely positive experience. Telecommuting offers potential employment to 10 million physically handicapped citizens. A Chicago-based organization taught computer programming to people with polio, cerebral palsy, spinal injury, and other afflictions. One man who attended the program had never had a job in his life, "and now he's a programmer for Walgreens."[41]

Computers do create jobs. Somebody does the designing, manufacturing, programming, selling, repairing, and other work. Of course, computers threaten jobs, too.

Demand for factory workers, clerical workers, and other unskilled and semiskilled laborers will drop as jobs become automated. At the same time, the demand for trained professionals, especially engineers and teachers, will rise sharply. There will be plenty of jobs as the twenty-first century progresses. But will there be enough skilled workers to fill those jobs? One key to a promising economic future is computer literacy.

4

Computer Literacy

AT ISSUE

In the 1960s, people who worked with computers were thought to be a little weird. In the 1970s, hobbyists who tinkered and learned about computers were considered to be nerds, and clever programmers were thought of as hackers. Most people had absolutely nothing to do with computers; they were quite content to be computer illiterate.

That attitude changed in the 1980s. People with computer knowledge were thought to be cool. Masses of people yearned to be computer literate and gain the skills they needed for a high-tech job. They also wanted their kids to have a computer education.

WHAT IS COMPUTER LITERACY?

The term "computer literacy" is one of many computer-related buzzterms, such as "information superhighway" and "interactive multimedia." Clifford Stoll writes, "Defining these [phrases] is like trying to nail Jell-O to the wall." [42]

Computer literacy can mean just learning to type on a keyboard or learning to use standard software applications such as WordPerfect or Excel. Some people expect users to be able to write simple computer programs or describe the functions of hardware components. Others think just playing computer games makes a person computer literate.

In Pennsylvania, the Academic Standards for Science and Technology include the goal of computer literacy. "The computer literacy standards were added in response to outcries

As personal computers became integrated into everyday life, the importance of teaching children to use them became obvious. Now, most schools in the United States have computer labs where students (and sometimes their parents) both learn to use and learn by using computers.

from the community. Business leaders, parents, and others want students to know something about computers. Consensus concerning exactly what students should know about computers is not so clear." [43]

Some advocates for computer literacy seem to think that computers will magically solve the shortcomings of the educational system. They believe "computer literacy will transform students into productive workers, informed citizens, and wise decision makers." [44] They also believe that computer science should be one of the basics taught to high school students, along with English, math, social studies, and science.

Due to this perceived need for computer literacy, the National Educational Technology Standards for Students (NETS) has attempted to set national objectives. The NETS project considers the computer literacy skills that every student should obtain to be:

1. Basic operations and concepts
2. Social, ethical, and human issues
3. Technology productivity tools
4. Technology communications tools
5. Technology research tools
6. Technology problem-solving and decision-making tools

COMPUTER EDUCATION FOR KIDS

"To live, learn, and work successfully in an increasingly complex and information-rich society, students must be able to use technology effectively," states the NETS program.[45] The objective is to develop standards to guide educational leaders in recognizing and addressing the essential conditions for effectively using technology in pre-kindergarten through twelfth grade education. NETS is a project of the International Society for Technology in Education (ISTE) and is supported by the National Aeronautics and Space Administration (NASA), the U.S. Department of Education, the Milken Exchange on Education Technology, and Apple Computer.

The public has also recognized the potential of technology, especially its ability to change education and improve student learning. As a result, more classrooms are being equipped with PCs. For example, in Maine, 18,000 seventh graders received a new laptop computer on the first day of school, courtesy of the Maine Learning and Technology Initiative.

The goal of the Maine initiative is to give young students the tools and training they will need to perform tomorrow's

Maine's Learning and Technology Initiative, a state-sponsored program, gave an Apple iBook to 18,000 seventh graders in the state in 2002. The program aims to prepare students for the tech-heavy job market of the future.

technical and intellect-based jobs. Maine Governor Angus King said the initiative, which amounts to 1.5 percent of the state's annual school budget, might be "the largest educational technology project in the history of North America." [46]

King also mentioned the impressive results of the program's pilot project. After just three months, he said, "disciplinary problems dropped 75 percent, absenteeism declined by two-thirds and, more importantly, the kids in the program developed a much more positive attitude about their schools and teachers." [47]

Providing access to computers at school is part of an effort to bridge the so-called digital divide between those who have computers and those who do not. Among households with school-age children, two-thirds have PCs. Educational programs aim to reach the other third.

For example, the Computers for Children program in Buffalo, New York, helps local schools meet their goals of computer literacy. The program holds summer and after-school classes in software instruction and PC repair. Students help to refurbish used computers, which are then donated to schools. Last year, 1,100 machines were refurbished and donated.

That more PCs are being placed in schools is certainly beneficial, but how much are they used? In some classrooms, the PCs are used for only a few minutes a day. Experts agree that there is still a need to educate teachers so they can expand the use of the PC in their classrooms.

Another concern is how the PCs are being used. Despite the popularity of special computer courses, such as keyboarding or word processing, most experts believe computer skills are best learned in the context of their practical application rather than when taught separately. All school subjects, from history to art to geometry, should teach and utilize PC skills.

Should all kids learn a programming language, such as BASIC, Pascal, or C? Some schools say yes and others say no, since specific languages can become obsolete within a year or two. "Training kids for entry-level, high-tech jobs makes no sense, given the present market," wrote journalist Joe Celko. "The problem in IT [information technology] is quality, not quantity. We are grinding out tons of people who have a certificate of some kind from a vendor but can't program their way out of an infinite loop."[48]

Another area of concern is that literacy is limited to certain skills. Who is tinkering with the guts of the PC? Who is showing kids the workmanship and artistry of the motherboard? A simple workshop that allows students to experience assembling and using basic computer components can be very educational. For example, examining the internal workings of a hard disk drive can help students visualize how data is stored and accessed on a disk. Hands-on exercises help demystify computers and make them more exciting.

Computers cannot provide a quick fix to education. They cannot substitute for good teachers and hard work. "Kids who walk up to my computer immediately ask what games I've got," writes Stoll in *Silicon Snake Oil*. "They're uniformly disappointed to hear that I don't have any. Computer games satisfy in ways that real life can never touch. You jump over the ravine. You blow up the alien. . . . If you're thwarted, why, just pull the plug."[49]

Stoll worries that time spent at the computer is replacing human interaction and exploration of the real world. A friend who bought his four-year-old son a computer told Stoll the boy would rather play computer games than play with other kids. Stoll writes: "This much is certain: no computer can teach what a walk through a pine forest feels like. Sensation has no substitute."[50]

PCs can even have harmful side effects. Obesity can result when children spend time sitting in front of a computer screen instead of running and playing. Repetitive strain injuries (RSI) can develop from using PCs and chairs designed for adults.

Playing violent computer games has been correlated with increased aggression in children in some studies. Too much time spent gazing at a monitor can harm a child's eyes. Teenagers may experience loneliness and depression as a result of excessive computer activity.

Most experts agree that there is a need to monitor children's computer usage. The amount of time children spend using

Early Calculating Devices

The quest for a device to perform tedious and boring calculations goes back many centuries. The abacus is considered the original mechanical counting device. It was in use over 5,000 years ago.

Mechanical devices with gears and levers came along in the 1600s. In 1642, French mathematician, philosopher, and mystic Blaise Pascal invented a counting-wheel device in order to help his father with his tax-collecting job. Pascal's counting-wheel design was used by all mechanical calculators until the mid-1960s, when electronics became more practical.

In the 1800s, British mathematician, astronomer, and inventor Charles Babbage imagined a steam-powered machine that could perform tedious mathematical calculations and provide accurate results. Babbage completed part of his Difference Engine in 1832, but never finished it. Meanwhile he had envisioned an Analytical Engine, a general-purpose computer that could add, subtract, multiply, and divide at a rate of 60 calculations per second. He intended for his machine to free people from repetitive and boring mental tasks, just as other new machines were freeing people from physical toil.

Babbage worked on the Analytical Engine until his death in 1871. Although he never saw it, a working model of the Difference Engine was completed in 1991 by London's Science Museum. The engine, built according to Babbage's original plans and drawings, stands six feet high, contains 400 parts, and weighs three tons.

Lady Ada Augusta Lovelace, who collaborated with Babbage, wrote articles about his ideas and added her own notes. She also wrote programs that could be used to instruct Babbage's engine to repeat certain operations. Many people regard her as the first programmer, and the U.S. Department of Defense named their standard programming language Ada in honor of Lady Lovelace.

computers and the types of activities they engage in are key factors in determining whether a PC has an overall positive or negative effect.

A COMPUTER DRIVING LICENSE

Many adults have resisted acquiring PC skills. Eventually some of them realized that they had to learn basic computer skills or

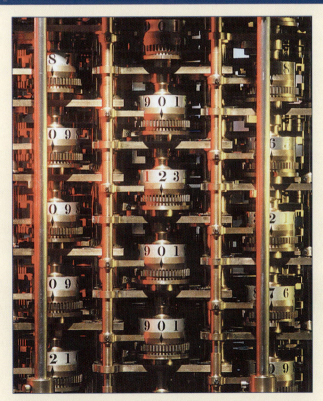

This front detail view is of Charles Babbage's Difference Engine, which calculated long and difficult calculations and eliminated human error.

risk being dismissed as relics of the past. Distressed employees worry that younger, more technically savvy workers will push them out of their positions. When these people see others using computers, they imagine that those people have access to all sorts of information. They don't understand computers, and some are scared of them.

"It will be hard to work in this country if you don't know how to use a computer," said Microsoft executive Dana Manciagli.[51] People who are comfortable using computers will have an easier time finding—and keeping—jobs.

Ford Motor Company, Delta Air Lines, and about a dozen other companies started programs that subsidize employee purchases of PCs. Under the three-year programs, Ford workers pay $5 a month and Delta employees pay $12 a month for a basic desktop model that they keep when the payments are completed. These companies believe that having employees who use computers at home will result in a more technologically literate workforce. They hope it might also boost employee loyalty and eventually reduce the costs of such processes as benefits enrollment by having employees use PCs instead of paper forms.

Employees have snapped up the machines. At Ford, 165,000 of 188,000 eligible workers in the United States and Philippines bought computers through the program. But whether workers will use their home PC to play games or to improve their computer skills is uncertain. Once again, the goal of computer literacy is vague.

There is no standard measurement of computer skills; the requirements vary from industry to industry and from employer to employer. Some universities require graduates to take a computer literacy exam to demonstrate their mastery of certain concepts and skills. Employers may require job seekers to take proficiency tests or to be certified in using specific software or hardware. Various levels of certification are available for Microsoft, Novell, IBM, and other companies' products.

Certification has become an important way to assess and validate specific computer skills and knowledge, but there is no standard for general computer literacy in the United States. However, in Europe, job seekers can show prospective employers their mastery of basic computer skills by flashing their computer driving license.

The European/International Computer Driving License (ECDL/ICDL), an internationally recognized computer literacy training and certification program, is becoming the global computer literacy standard. The ICDL is now available in the United States through courses and testing at some colleges and training centers. Soon, everyone might need a computer driving license.

There are many ways to prepare for the ICDL test or just to improve your PC skills. Continuing-education classes, educational software applications, and distance-learning courses offer computer training. Computer user groups, which are essentially self-help support groups for computer users, are also available for all makes of computers and all kinds of software.

Another way to get help is to ask young people to become trainers. Retiree Martin Schrader overcame his computer illiteracy with help from his son. He now says: "My advice to retirees who want to get back into some kind of action: Learn, baby, learn, no matter whether you are in finance, human resources, retail . . . It's not hard. It's very rewarding and whatever you do, you'll do it better with these modern tools. Oh, yes, these computers are great toys, too."[52]

Community programs also offer access to PCs and training. Community Technology Centers' Network (CTCNet) offers low-income children, youths, and adults opportunities to learn to use computers and other technologies, such as the Internet. National studies confirm that low-income and minority Americans lag behind other groups in ownership of computers. It is difficult to document the impact of these inequities, but community leaders are convinced that a lack of

One technique businesses use to keep their employees up-to-date on computer skills is to sponsor classes in using new software, like the one shown in session here. Many community colleges, public high schools, and community programs also offer classes that help users become computer literate.

computer skills is a handicap in the job market. Community Technology Centers' Network found that:

> Mastering the computer provided a tremendous sense of individual satisfaction and accomplishment. Often the people who use these community centers have not had good experiences in school, and having a different kind of learning experience at the center has a dramatic impact on their self-images. Toni Stone calls this the "I can!" phenomenon, and points out that it is self-reinforcing: one success leads to another. "You never know where the first 'I can' will lead, whether you are homeless, illiterate, disenfranchised, or even an otherwise educated person." [53]

Computer literacy seems to have the power to improve self-esteem, increase job prospects, and change a person's outlook on life. "Mastering this complex, modern tool is a powerful, affirmative experience—one that offers challenge as well as hope."[54]

5

Technophobia and Technostress

AT ISSUE

As people spend more time at their computers, they report a number of unwelcome side effects. High stress levels and mental fatigue are common complaints, as are vision problems, headaches, backaches, neck pain, and wrist and hand problems. In 1990, RSI (Repetitive Strain Injury) was the leading workplace illness, accounting for 185,000 workmen's compensation claims at an estimated cost of $20 billion.

It is clear that working at a computer can be hazardous to your health. But there is nothing inherent in the computer that causes pain and discomfort. Virtually all of the problems can be reduced or eliminated by quite low-tech, low-cost measures.

FEAR OF THE COMPUTER

For some individuals, the PC has become a comfortable, interesting, and nonthreatening companion. Hackers prefer it to other human beings. Some kids can't get enough time behind a keyboard. Others find themselves bored, and many feel a deep sense of frustration when working with PCs. In work settings, two-thirds of employees are "hesitant" about technology.

"The personal computer is perhaps the most frustrating technology ever," says Donald A. Norman, author of *The Invisible Computer* and a former vice president at Apple Computer. "It should be quiet, invisible, unobtrusive, but it is too visible, too demanding." [55]

Along with the benefits of PCs have come some negative side effects, including physical complaints, stress, and mental fatigue. Because few users understand how computers work, they often experience frustration when computer problems arise. In the workplace, two-thirds of employees are said to be "hesitant" about technology.

The frustration makes some users want to smash the screen, kick their desks, or throw the mouse out the window. Low-tech people caught in high-tech times experience what is called "technophobia," the fear of technology. Feelings of paranoia and dislocation boil inside them.

Back in 1994, the complaints included "the complex and creaky DOS operating system, the painful process of configuring or expanding an IBM-compatible PC, the miserable state of instruction manuals, and the difficulty of getting a lot of today's multimedia software to run properly on your computer without disabling other programs."[56] It is not surprising that in a 1995 survey of American PC users, 59 percent admitted getting angry at PCs within the previous year.

Some adults, particularly older ones, have trouble adjusting to the information age. They grew up before computers entered the mainstream. One user describes trying to help his 80-year-old relative, Frank:

> The most interesting thing I noticed about the way he approached the computer was the fear of making a mistake and causing irreparable damage. As he typed on the keyboard, letter by letter, he was OK until he inevitably hit the wrong key.
>
> After he calmed down, I realised that the mental model Frank was working from was that of a typewriter. Make a mistake, and you use white-out or start again.[57]

Like Frank, many first-time users worry that making one mistake will ruin everything. In fact, most programs make it very hard to lose your work. A quick click on the Undo button is usually all it takes to recover. The designers of PCs do try to make them user-friendly.

Computers have those cute little icons and make happy sounds when you execute commands. The screen glows warmly and the mouse clicks softly. Then, a minute later, it freezes and throws an impersonal message on the screen: "This program has performed an illegal operation and will be shut down."

Sometimes, the people who come to help the novices

interpret these error messages are not very friendly, either. One mother wrote to the *Wall Street Journal* to complain about her kids' help. They reprogrammed her Macintosh so that when she makes an error, it says, "You silly billy!" She says this is "not what I want to hear as my frustration mounts."[58] Of course, she cannot figure out how to remove the message.

Why do PCs have so many frustrating quirks? Software programs are becoming easier to use, but at the same time, they are offering more choices. Author Donald A. Norman feels that "the industry has succumbed to a technology fever, to the disease of featuritis, to pushing new technologies at the customer faster than even the most compliant customer can absorb."[59] The added complexity does not mean users have more power, it means they're overloaded with a maze of possibilities. It means they spend more time learning how to use the software, and less time actually using it productively.

One survey found that people who received excellent to good training had positive attitudes about computers whereas those who received fair to terrible training had more negative reactions. It seems that not being taught computer skills properly can increase stress levels for users, who may not feel comfortable with a certain software application or computers in general.

Of course, as soon as users get comfortable with their current hardware and software, they need to move on to new ones. "'Migrating' to a new PC can range from hellish to exhilarating, depending largely on how well you prepare."[60]

As for PC hardware, it was initially built without any thought being given to the user's comfort. That is nothing new. The arrangement of keys on the typewriter, the predecessor of the computer keyboard, was actually designed to reduce typing speed because fast typists were jamming the keys. Thus, the most commonly used keys are as far apart as possible. This

QWERTY arrangement, named for the top line of keys, is neither productive nor healthy, but it remains the standard.

Computer consultant Howard Moscowitz offers some techniques for working with computerphobes. One way he makes the PC seem less intimidating is by opening up the system unit case and showing people what is inside. "One of my most effective strategies," Moscowitz says, "is to disarm people through humor; sometimes I even use such a silly ploy as asking them to hug the computer."[61] This gets people to personalize the computer.

Finding something that a PC does that makes life easier or more fun can help people warm up to the machine, too. Many older people who are interested in genealogy use the PC to research and record their family's roots. Others use it to write poetry or play solitaire. Using the computer in these ways makes the computer seem less imposing and more helpful.

It also helps if people realize that almost everybody has technostress—meaning that they feel stressed out by technology. "High-tech living has thrust us all into an inescapable state of terror and revulsion that is wearing us out," writes conservative columnist Florence King. "The ease of computers and the instantaneous reach of fax machines are all well and good, but they are so sensitive and complex that each time we turn them on we brace ourselves, heads cocked like paranoid maniacs, listening for the various hums and buzzes and ding-a-lings to judge whether they sound the way they usually do. If they don't, we go into instant nervous prostration and give vent to the *cri de coeur* of the helpless, cringing souls we have become: 'It's doing something different!'"[62]

Another cause of stress for PC users is that time has been compressed. PCs can do things faster, and that has increased people's expectations of how quickly they should be able to accomplish tasks. People measure their performance against an impossible yardstick. Then, they get impatient while waiting one minute for the PC to boot.

PCs have also shifted the focus from doing a single task to doing multiple tasks. But a PC being able to do many things at the same time does not mean that people can work that way. Simultaneously doing a number of tasks—for example, sending e-mail, browsing the Web, and doing word processing—is multitasking gone mad. Just because the software can run 37 windows at once does not mean that doing so is expedient or beneficial.

Some people have trouble sitting down and being thoughtful. They suffer from information overload. Futurist Paul Saffo says, "It used to be considered a status symbol to carry a laptop computer on a plane. Now anyone who has one is clearly a working dweeb who can't get the time to relax. Carrying one means you're on someone's electronic leash."[63]

The constant presence of a PC can definitely cause stress. "My husband and I have this ongoing fight over whether the computer is allowed in the bedroom," says Brenda Laurel, founder of computer-game company *Purple Moon.* The bedroom should be a "no-computing zone," she says, because she wants a vacation from the PC when she goes to bed. "At some point, you have to live without technology. Almost every day you see someone using technology in a way that intrudes on someone else's life."[64]

Many clerical workers in the United States are monitored to some extent by a computer. Their work computer can measure and record how many times they hit the keys, how many errors they make, and when and how often they take breaks. Workers say that the situation increases stress and negative attitudes toward work and computers. James Irvine, a workers' union official, told the *Washington Post* in 1984, "The whole world is changing and getting so computerized and dehumanized. It's important to have human beings, they ought not to be treated like an appendage to the goddamn computer."[65]

But author Ray Bradbury commented on attitudes of that sort in 1985. He wrote:

> I see nothing but good coming from computers. When they first appeared on the scene, people were saying, "Oh my God, I'm so afraid." I hate people like that—I call them neo-Luddites. They're just like the people who used to run into factories and beat up on machines with sticks. They say, "The computer will know everything about you." My response is, so what? I've got nothing to hide.[66]

PHYSICAL PROBLEMS

A person's physical well-being is dependent on their emotional and mental states as well as their physical condition. Having a positive attitude toward technology, like Bradbury's, can yield positive physical results. Conversely, approaching the computer with a negative attitude can influence your body posture and breathing. Hostility and bad feelings toward the PC lie at the root of many of the physical health problems attributed to computer use.

Computer-related health problems result in medical expenses and lost productivity to the tune of about $10,000 per afflicted employee. "They can top $100,000 if legal liability for industrial injury becomes an issue," say Dr. Ronald Harwin and Colin Haynes, authors of *Healthy Computing*.[67] That adds up to a lot of pain and disability caused by the seemingly innocuous activity of using a computer.

There are a variety of sources of these problems. Something as simple as the noise a computer makes may be harmful to those who are sensitive to it, even if they are not conscious of hearing the noise. Laser printers and computer cooling fans are also major suspects.

Like many other appliances, most computers and monitors emit low-level electromagnetic fields. This radiation may pose

Repetitive strain injury, or RSI, is one of the most common injuries among computer users. Long hours spent at improperly set up workstations can produce pain, numbness, inflammation, and nerve compression, particularly in the wrists and hands. Supportive devices, such as these wraps for the wrists, can help to alleviate pain and prevent additional injury.

health problems. In the late 1970s, reports circulated about computer monitors causing cataracts, miscarriages, and birth defects. The reports were never verified, but the scare prompted a change in the technology.

Cathode-ray tube (CRT) monitors made since the early

1980s emit very low levels of radiation. The strongest fields are emitted from the monitors' sides and back. An alternative is to use an LCD (liquid crystal display) screen, which emits negligible radiation.

Monitors are also responsible for many eye problems. In 1991, eight million workers complained to their eye doctors about monitor-related problems. Surveys repeatedly show that workers rank eyestrain as their most serious concern among workplace hazards.

"Eyes are the softest, most vulnerable part of the body that is exposed to the external world. They are your most sensitive interface with your environment," explain Harwin and Haynes.[68] The condition of a person's eyes can profoundly affect their psychological and emotional well-being.

Human eyes were made for seeing things at a distance, not for doing close-up work for hours at a time. When working at a PC, users should not keep their eyes fixed on the screen. They need to let their eyes relax, perhaps by simply looking out the window from time to time.

Bright lights or sunlight coming through a window can cause glare on the monitor screen. Removing overhead lights, installing a window shade, or using an antiglare filter on the monitor may help. Software options that let users change the text size, style, and color might reduce eyestrain.

Skin rashes also might be linked to working for long hours at a PC. A Swedish study found a possible explanation: "Certain computer monitors emit a chemical that can cause allergic reactions."[69] The scientists discovered that the plastic case on some computer monitors contained up to 10 percent triphenyl phosphate by weight. The heat generated by the monitor caused the compound to start evaporating. Soon, a small but measurable amount of the pollutant tainted the air.

The good news was that after 10 days of operation, the emissions fell to one-third of the initial amount. After the equivalent of two years of use, the rate of emissions was just 10 percent of what it had been.

Perhaps the biggest concern for computer users is the hazard of RSI (Repetitive Strain Injury) to the hands and arms. These injuries occur from repeated physical movements and can damage tendons, nerves, muscles, and other soft body tissues. Musicians, knitters, and computer users are among the people susceptible to RSI because of the repetitive tasks they perform.

In 1992, RSIs were reported to be causing more than 16 million lost workdays each year. That total does not include professionals and self-employed people who tend to be heavy computer users, but whose health problems are not well documented.

RSI threatens to become an even bigger problem. More people are using computers and experienced users are spending more time at computers. A long period of typing on the flat keyboard of a laptop is especially harmful. So are hours spent clutching, dragging, and clicking a mouse. These kind of repeated movements are likely to increase the number of cases of RSI.

"This can be a serious and very painful condition that is far easier to prevent than to cure once contracted, and can occur even in young, physically fit individuals," says Paul Marxhausen, who runs the Computer Related Repetitive Strain Injury Web site. "It is not uncommon for people to have to leave computer-dependent careers as a result, or even to be permanently disabled and unable to perform tasks such as driving or dressing themselves." [70]

RSI is a general term for a number of conditions, including tendinitis, bursitis, thoracic outlet syndrome, trigger finger/thumb, and myofascial pain syndrome. All of these are serious and can cause great pain, but the most publicized condition is carpal tunnel syndrome (CTS). The symptoms of CTS are pain, numbness, and burning or tingling sensations in the fingers and hands. It results from compression of the median nerve in the carpal tunnel, located in the wrist by the thumb. Swelling of the tissues in this area is common among computer users because of their use of keyboards, mice, and trackballs.

In the 1990s, CTS surgery became the surgery of the decade. Unfortunately, the operation is expensive and carries no guarantee of improvement. Patients lose hand movement for two to six months after surgery, and often, after they return to normal activity, the pain returns due to a buildup of scar tissue.

ERGONOMICS

Anyone who regularly experiences the symptoms of RSI when working at the computer should see a doctor right away. Early diagnosis can limit the damage and prevent a lot of pain and serious injury. The best treatment plan varies immensely according to the severity of the injury and the individual.

In general, simply using gadgets like wrist splints, palm rests, or split keyboards will not permit the user go back to work at full speed. Even people who undergo successful CTS surgery may still experience pain. For computer users to recover from RSI and prevent it and other health problems from developing, three things are necessary: (1) using correct keyboarding technique and posture, (2) using the right equipment setup, and (3) having good work habits.

These concepts are part of ergonomics, the science that coordinates office design with workers' needs. The basic idea of ergonomics is that workers should not fit themselves into the workspace, but rather that the job environment should be fitted to the size and needs of each worker. Creating a healthy workspace means choosing equipment that is ergonomically designed.

The most important items are the chair and desk, which must be positioned at the proper height. The desk and chair should be situated so that the wrists are straight and level when placed on the keyboard, not bent down or up. The monitor should be at or just slightly below eye level. The feet should be flat on the floor, with the knees lower than the buttocks.

While an ergonomically designed workspace is important, users also need to look at themselves and their habits, including posture, repetition, and fatigue. One of the best ways to prevent injury is to limit computer use. No amount of good technique or exercise is going to help if a user simply spends too much time at the PC. Users should especially avoid games, which often involve very intense use of the keyboard, mouse, or trackball.

Because computer equipment is not designed or set up correctly for their body size, kids are at risk for computer-related health problems, too. The workspace must be tailored to their size and they must develop healthy usage habits. Otherwise, children are risking injury. Younger bodies are more forgiving than those of adults, but serious problems could appear in the future.

Dr. Alan Hedge, a design and environmental analysis professor at Cornell University, thinks there is a three- to five-year window to prevent injuries. "But there's an ostrich mentality—we may have to wait until children become injured and start suing to make real changes," he said.[71]

Most experts agree that kids can sit for about an hour in adult-sized chairs without any discomfort. For longer periods of sitting, it's recommended that the chair fit the child better. This may mean raising the seat and pushing the lumbar support forward, or simply placing a pillow in the seat and behind the lower back. If a child's feet dangle, a footrest or box can be used to support them.

Dr. Hedge said: "When they mass-computerize schools—a computer on every desk—they need to make sure it's the right desk. Otherwise, you'll see the same problems we see in adults."[72] But it is difficult for schools to make computer workstations ergonomically correct because children share computers. Each child may need a special setup. Also, money for furniture usually is not a priority in an already tight school budget.

Laptop computers were not designed to be used all the time, but for many users, a laptop is their only PC. Because the monitor and keyboard are attached, putting the computer in a comfortable position for the arms, like on the lap, as seen here, makes the user tilt his or her head forward to see the screen. Neck and back pain are often the result.

Ergonomics consultant Dr. Inger Williams thinks, "The problem is that schools are not aware that these are concerns and they're worried that it will cost money to introduce ergonomics." But pillows and boxes can solve many problems. Once the concepts behind ergonomics are explained, however, "Children really resonate with ways to make themselves more comfortable," Dr. Williams says.[73]

Laptop computers are also creating health problems. These small computers were not designed to be a user's primary computer; they were designed with portability in mind, not user comfort. Now, however, people use them all day.

The keyboard and monitor are attached in a laptop, so their positions cannot be adjusted. The result is a compromise. If users put the PC in their lap, they have a comfortable arm position but have to tilt their neck forward to view the screen. If they raise the screen, their hands must reach too high to type comfortably.

The keyboard itself is another problem. The keys on a laptop's keyboard typically are smaller than on traditional keyboards, which may cause finger cramping or hand pain. In place of a mouse, laptops use a little pointing stick, a trackball, or a flat touch pad. These devices adequately control cursor movement, but the ergonomics of most are poorly designed.

Companies make a variety of laptop accessories to remedy some of these problems. Document holders, external keyboards, and detachable screens are available. But a better way for laptop users to avoid some aches and pains is to take more frequent breaks. Switching position every 30 minutes is recommended.

Using a computer should be an enjoyable, pain-free experience. By properly adjusting equipment, learning healthy techniques, practicing good work habits, and letting go of negative feelings toward the computer, users can eliminate many of the pains and discomforts associated with computer use.

6

Hackers, Crackers, Pirates, and Viruses

AT ISSUE

In 1982, a specialist in computer abuse said, "Nobody seems to know exactly what computer crime is, how much of it there is, and whether it is increasing or decreasing. We do know that computers are changing the nature of business crime significantly."[74]

The PC revolution presented legal authorities, programmers, and computer users with some new ethical and legal issues. Software companies had to fend off pirates. Network administrators dealt with hackers and crackers. Personal computer users learned how to protect themselves from Trojan horses and viruses.

GOOD HACKERS

In his book *Hackers*, Steven Levy defined his subject as "those computer programmers and designers who regard computing as the most important thing in the world."[75] He viewed them as a new breed of American hero, "whiz kids whose irreverence, idealism, and sheer genius changed the world."[76]

Some people used the word hacker to mean "nerdy social outcasts or 'unprofessional' programmers who wrote dirty, 'nonstandard' computer code." But Levy saw them differently. "Beneath their often unimposing exteriors, they were adventurers, visionaries, risk-takers, artists . . . and the ones who most clearly saw why the computer was a truly revolutionary tool."[77]

The first hackers were veterans of the Massachusetts

Hackers, shown here at the 2002 H2K2 hacker conference in New York City, see themselves as technically savvy, irreverent, and adventuresome users exploring the information frontier. To the general public, however, the term "hacker" quickly took on a more sinister, antisocial connotation.

Institute of Technology's Tech Model Railroad Club. When club members discovered that one of MIT's minicomputers was available late at night, they started writing programs instead of toying with model railroads. Joseph Weizenbaum, an MIT professor, described these hackers as "bright young men of disheveled appearance, often with sunken glowing eyes." He said, "Their rumpled clothes, their unwashed and unshaven faces, and their uncombed hair all testify that they are oblivious to their bodies and to the world in which they move. They exist . . . only through and for the computers. These are computer bums."[78]

Hackers do not view themselves as bums. To them, the term

"hacker" is a complimentary classification, indicating respect and recognition of a person's computer savvy. True hackers consider the term a badge of honor rather than an insult. The hacker community even devised the Hacker Ethic, which stated:

- Access to computers should be unlimited and total.

- All information should be free.

- Mistrust authority; promote decentralization.

- Hackers should be judged by their hacking, not bogus criteria such as degrees, age, race, or position.

- You can create art and beauty on a computer.

- Computers can change your life for the better.[79]

When the PC became popular, the hacker movement became more widespread and the term hacker began to lose its original meaning. A "good hack" had been a good thing, but then hacking began to connote using computers for antisocial or illegal purposes. "Cracker" is the correct slang term for a person who gains unauthorized access to a computer system with the intent to destroy data, steal software, shut down hardware, manipulate data, or do some kind of harm. Few people distinguish between nondestructive hackers and crackers; they refer to all of them as hackers.

Conflicts between hackers and law enforcement grew in the early 1990s. "These hackers are explorers, not criminals or vandals," said John Barlow, lyricist for the Grateful Dead rock band and cofounder of the Electronic Frontier Foundation (EFF). "They're exploring a new information frontier. It's a reincarnation of what happened with the settling of the Old West, only in the computer sphere."[80]

Government officials and law enforcement agencies did not see it that way. "Many computer hacker suspects are no longer misguided teenagers mischievously playing games with their computers in the bedroom," said a Secret Service agent. "We will continue to investigate aggressively those crimes which threaten to disrupt our nation's business and government services." [81]

Despite the government's opinion of hackers, some businesses still thought of them as benevolent computer users. They felt hackers could be helpful by finding security flaws and program errors. In 1995, Netscape Communications offered cash rewards to people who found flaws in the test version of its newest browser software.

Time magazine explained: "Under the so-called Bugs Bounty program, the first person to identify a 'significant' security flaw wins $1,000. Lesser bugs earn smaller prizes ranging from $40 sweatshirts to $12 coffee mugs. The idea, explains a company spokesperson, is to get hackers to hack when it will do Netscape some good—before the product is officially released." [82]

The Bugs Bounty program was Netscape's response to the embarrassment the company had suffered the month before. Hackers had cracked its browser security code and Netscape had had to quickly release a new version of the software. The breach had followed the company's promise that they would make the Internet safe for business.

In the end, Netscape was even more embarrassed. They paid more than 20 people $1,000 for new bug reports. Then, Danish programmer Christian Orellana found a security bug in Netscape's browser and asked for more money. The press labeled him a high-tech extortionist, but other observers thought Netscape bore some responsibility for promoting the scheme in the first place.

SOFTWARE SABOTAGE

Destruction unleashed on PCs often arrives in the form of a virus. Computer "bugs" (programming errors) had been around for years. Then, in 1983, Fred Cohen at the University of

The proliferation of both computers and computer viruses made a means of protecting systems vital to the PC industry. Antivirus software, such as McAfee VirusScan and Norton AntiVirus, must be constantly updated to keep pace with the malicious programs that threaten users worldwide.

Southern California coined the term "computer virus." A virus, Cohen said, was "a program that can infect another program and replicate itself inside its host."[83] Without the user's knowledge, the virus enters the computer and goes into action when the affected program is run, infecting files, applications, or the boot sector, which contains the code that starts the computer when you turn it on.

In 1992, the first notorious virus spread through personal computers. The Michelangelo virus made use of other trickery referred to as a Trojan horse and a time bomb. A Trojan horse, named for the legendary Greek subterfuge, is malicious code hidden in a program that performs a useful task; unwitting users activate the destructive code when they run the program. Some Trojan horses are set to attack in response to a particular event, such as the entering of a special code in a database field. If the destruction is triggered by a time-related event, it is called a time bomb.

The Michelangelo virus entered a PC through an infected disk and replaced the machine's boot sector. The virus performed the normal start-up functions, but it also duplicated itself and infected any floppy disks put in the disk drive. Meanwhile, the time bomb was waiting for Michelangelo's birthday, March 6. If the boot sector was activated on that day, the data files on the PC's hard disk were destroyed.

Media reports estimated that the Michelangelo virus might damage 5 million computers. But because the public had been warned of the impending time bomb, only about 2,000 machines suffered serious losses. Still, the scare raised the public's awareness of computer viruses.

As more viruses circulated and more people bought computers, a large market developed for software that protected PCs from viruses. John McAfee wrote the best antivirus software program—and gave it away for free. McAfee had an astonishing plan in mind. He would give away the software, build a large and loyal customer base, and then "throw the switch" and start charging for product upgrades.

Unlike traditional software production, in which companies package and shrink-wrap software disks and then ship them to retailers and customers, McAfee delivered his software via dial-up bulletin boards. Customers could download the product and distribute it to friends and coworkers. "It didn't cost me a dime," says McAfee. "The magic of software is that once it's developed, duplication is instantaneous and has zero cost."[84]

McAfee made his customers happy and soon his antivirus software became the standard. Then he started charging customers, and most stayed on board. McAfee said:

> They're used to your interface, your documentation, your way of viewing the world through your product. When you start charging for updates, it's very easy for them to rationalize paying for it rather than going to a competitor. I've been using the software for two years. I like it. They've done it for free. I like the company. Sure. I'll give these guys five bucks a month.[85]

McAfee Associates was extremely successful. By July 1998, the company was worth $3 billion.

SOFTWARE PIRACY

Copyright is the exclusive legal right of the author of a creative work to control the copying of that work. Historically, authors could copyright works consisting of text, artwork, music, audiovisual materials, and sound recordings. When the Computer Software Copyright Act was passed in 1980, it declared programs to be "literary works," thus giving them protection by copyright laws.

Software piracy is the illegal duplication of copyrighted computer software. Freeware (also called public domain software), which is freely distributed by disk or via the Internet, and shareware, which is free for the trying and then requires payment for continued use, can legally be copied and shared. Commercial software cannot.

Piracy of copyrighted works has always been simple to perform. Books were photocopied and cassette tapes duplicated, but software gave piracy a new twist. The PC made it possible to make copies that were identical to the original software. The process was fast, easy, and very inexpensive.

Software piracy became an issue as soon as the first personal computers were sold. In the early days of the Altair 8800, pirated paper-tape copies of BASIC were somehow obtained and shared among users. Hundreds more copies of the language were in use than Microsoft had sold. A furious Bill Gates wrote a letter that was published in the Home Brew Computer Club Newsletter.

Gates later recalled: "I wrote a widely disseminated 'Open Letter to Hobbyists,' asking the early users of personal computers to stop stealing our software so that we could make money that would let us build more software." [86] It did not convince the hobbyists, however. They knew they had to pay for hardware, but they preferred to "borrow" software.

David Bunnell recalled: "It created a firestorm. And the basic problem was, in my opinion anyway, that in the letter [Gates] said, 'You are all thieves.' And the problem with that was that they weren't actually *all* thieves, just most of them. It was really the first time that, at least in the personal computing industry, . . . software piracy became a big issue." [87]

Users felt that making one copy for their own use was not hurting the big software companies. But it was, as Paul Allen, cofounder of Microsoft, explains: "We put a lot of blood, sweat, and tears into making that BASIC; a lot of late nights, you know, a lot of hard work went into those." [88] In essence, Microsoft's entire business was those paper tapes, and the royalty for each one mattered.

In 1994, it was estimated that half of all the software in the world was pirated. This amounted to lost income of about $10 million. Software companies pointed out that prices for their products remained high because they had to make up for the sales lost to piracy.

Critics argued that the loss estimate was based on the assumption that every illegal copy would have been purchased. They claimed that many people simply would not use the software product if they had to pay for it. Some critics also pointed out that it was not the individual user who was responsible for the majority of losses; corporate America duplicated PC software for employees without paying for multiple copies.

For a while, many software companies used physical copy protection schemes so that writing a perfect copy of a disk was impossible. "Copy-protection used to be included on *everything*, from games to business software, simply because software was extremely expensive back then, and a couple hundred copies of a program could actually make or break a software company (no, I'm not kidding)," wrote a hacker who calls himself Trixter.[89]

A typical scheme was to hide a piece of data in a specially formatted track. When the program started, it checked for that

Keeping Your Computer Safe

Computer security is very important. Viruses, hardware failures, power failures, and other disasters can destroy data, equipment, and productivity.

Theft of computers, especially laptops, is one problem. Dropping the computer, fire, flood, and spilling liquids on the keyboard can also damage the hardware.

A power fluctuation through the power line — a spike or surge — could fry the insides of your PC. To avoid this disaster, plug your PC into a surge protector or surge suppressor that is then plugged into an outlet.

While hardware and software are relatively easy to replace, data often is irreplaceable. Make backup copies of important files and keep them in a safe place.

Handle diskettes with care. Store them in boxes. Remember that disks do not last forever, so you may want to copy them onto tape or CD-ROM.

Be very careful about sharing floppy disks. Viruses can be transferred on disks. Be aware that freeware, shareware, and pirated software are known to be sources of viruses.

Regularly use and update virus protection software. It will catch the majority of known threats.

special piece of data and if it did not find it, the program stopped. Thus, the program would run only from its original disk.

Steven Levy explains how hackers felt about the situation:

> Copy protection was like some authority figure telling you not to go into a safe which contains machine-language goodies . . . things you absolutely need to improve your programs, your life, and the world at large. Copy-protect was a fascist goon saying Hands Off. As a matter of principle, if nothing else, copy-protected disks must therefore be "broken."[90]

Companies tried all kinds of schemes to prevent copying, but hackers learned the schemes and how to get around them. Most tricks involved the way the DOS command DISKCOPY worked. For example, the company might format a disk as single-sided, but then store the special bit of information required for the program to run on the second side. DISKCOPY would determine that the disk was single-sided and only copy the first side. The duplicate disk then would fail to run the program.

One of the most effective methods of copy protection was to physically damage the floppy disk. Trixter explains: "Using a laser, it was possible to burn a small hole in the disk surface, and then all the program had to do was check to see if there was a read error in that particular sector, and if so, continue running the program. If you turned the disk surface manually by grabbing the inside ring, *you could actually see a tiny hole in the disk surface!*"[91]

Companies eventually stopped using copy protection since the schemes often reduced the reliability of the software disks and hackers generally found ways around it anyway. Copying became easier, but was still illegal. Today, most users understand that software is protected by copyright. However, software piracy is still a major issue in trading with other countries that do not have, or do not enforce, copyright laws.

Apple Computer had to battle the musical group the Beatles when it first included CD-ROM drives in its machines to play music. The company had agreed not to go into the "music business" to settle an earlier lawsuit with the Beatles' record label, Apple Records, over the use of the name Apple. It was argued that the CD-ROM drives made Apple Computer part of the music business. Apple Computer settled the suit by paying the Beatles $30 million so users can listen to music while they work or play.

In China, about 95 percent of all new software installations are pirated; in Vietnam, the piracy rate is 97 percent.

The Internet changed the nature of the situation again, since copies could be distributed quickly and repeatedly online. The Business Software Alliance estimated that in the year 2000, more than $11 billion was lost due to software piracy. Part of the solution is to educate people about copyright laws regarding online material. For example, things are copyrighted the moment they are written, and no copyright notice is required for their protection.

In a 1986 court case, copyright protection was extended to the "look and feel" of a computer program, the commands used to operate it, and the way it looks on the screen. Lotus Development Corporation won a suit against Paperback software because their program used the same commands as Lotus 1-2-3.

Apple sued Microsoft over the Windows user interface, which Apple said looked a lot like their interface. Then Xerox sued Apple, claiming that the GUI that Apple used was actually their invention. Xerox lost that lawsuit, and Apple lost the suit against Microsoft.

Apple seemed to have more than its share of lawsuits. Shortly after it was founded, Apple Computer was involved in a trademark dispute with the world-famous rock band the Beatles. The band recorded on their own label, Apple Records, and claimed the computer company had no right to use the name Apple. To settle the dispute, Apple Computer agreed not to go into the music business.

Years later, when the Macintosh included a CD-ROM drive to play music, the Beatles brought back the lawsuit. Apple Computer settled by paying about $30 million dollars to use the Apple name. In summarizing the legal battle, the Beatles' lawyer borrowed one of the band's song lyrics, saying: "It's been a long and winding road."[92]

7

The Incredible Shrinking PC

AT ISSUE

Technological advancements have allowed engineers to build smaller computers. Pocket PCs and laptop PCs are light and can operate on battery power. Personal digital assistants (PDAs) or handheld PCs are very small and generally support a specific application. There are even "hands-free" PCs that can be worn on a belt or shoulder strap and activated by voice interface.

Ray Kurzweil, an award-winning inventor and high-tech entrepreneur, predicts that by 2009, computers will be embedded in our clothes. By 2019, he thinks computers will be largely invisible and "embedded everywhere—in walls, tables, chairs, desks, clothing, jewelry, and bodies."[93]

THE PC ERA

Many industry observers say the PC revolution is over, that the PC has had its period of fame and fervor. The next generation of computing, they predict, shall be called "the post-PC era."[94]

Longtime industry analyst John Dvorak thinks that notion is wrong. "The PC era is standalone and forever, and we should get used to it," he says. "The personal computer is the most successful consumer electronics device in history, and yet people seem unable to think of it as anything other than a passing fad . . . they've been on the market for 24 years! Do people still think they're going to go away?"[95]

Dvorak agrees that there will be changes. He thinks computers will be embedded in all kinds of gadgets, but the

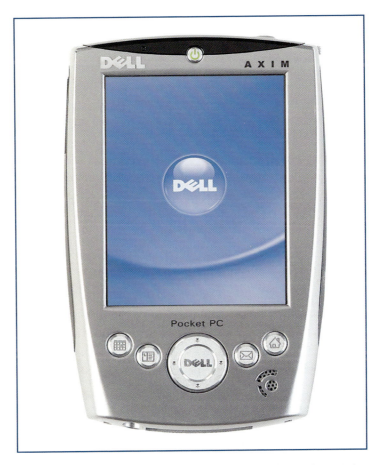

Handheld computers like Dell's Pocket PC have resulted from technological advancements that permit much smaller components to be used. Most handheld PCs support a specific application, such as a Web browser.

general-purpose PC will prevail. The PC's success, he says, has been because of its versatility, not in spite of it. But other industry observers think the "one size fits all" PC will fade away and be replaced by a multitude of devices designed to handle specific user tasks.

The PC era does show signs of aging and of slowing down. Perhaps the glory days are over. The market for PCs in North

America has stalled at between 55 and 65 percent of households. It is now a replacement market, and a tough sell. For the first time in history, the U.S. PC market experienced negative growth in 2001. Of course, that still equates to estimated sales of 45.3 million units.

Worldwide shipments of PCs also declined to about 124 million computers in 2001. But there are still many untapped markets. PC use in countries such as China, India, Vietnam, and Thailand is growing at a rate of more than 20 percent a year. Other opportunities for sales exist in Europe, where the ownership rate is only 27 percent, and in Asia, where the rate is 35 percent. In the United States, only about 15 percent of households own multiple PCs; the growth potential is substantial.

Improvements in PCs might expand the market, too. For example, the NEC PowerMate eco is designed to be environmentally friendly. It consumes very little power, which not only saves electricity but also means the system generates very little heat and does not need a fan. This reduces the noise level to no more than 20 decibels, which is barely whisper loud. The motherboard on the PowerMate uses lead-free solder, the monitor contains no boron, and the case is made from 100 percent recyclable plastic.

Newer laptop computers have improved battery life and are lighter. The IBM ThinkPad X30 weighs only 3.7 pounds and has a six-cell battery that lasts more than four and a half hours. By attaching an optional battery, the power duration can be increased to almost nine hours—a full business day.

Continued improvements in price performance should attract more buyers, too. Currently, the average consumer pays less for more computing power and storage capacity than ever before. In addition to setting low sticker prices, many companies are trying to lure customers by hyping the PC's ability to connect to MP3 (MPEG Audio Layer 3) players, digital cameras, and other gadgets.

IBM's ThinkPads are among a new generation of lighter laptop PCs that have longer-lasting batteries. The x30 model, which has a six-cell battery that will power the machine for more than four and a half hours, weighs only 3.7 pounds.

THE EXTENDED PC

Intel sees the future as the era of "the extended PC." They envision a desktop PC with a Pentium 4 processor as the control center and many intelligent tools to interface with it. These tools could include PDAs, digital cameras, digital video recorders, MP3 players, DVD (digital versatile disc) drives, CD-RW (compact disc-rewritable) drives, scanners, e-books, PC-enhanced toys, and cell phones. The devices will share content with the PC, including music, video, and photo files.

As technology develops over the next few years, users will be

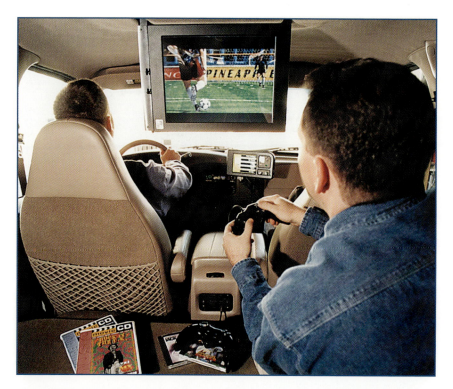

Intel officials predict that in the next few years, the "extended PC" will become commonplace. Many different electronic devices will interface with individuals' PCs, which will always be "on" and available in everything from cars to telephones.

able to connect a PC to networks in any airport, hotel, or coffee shop. Software will pull digital content and all computing devices together to deliver an integrated experience. Computers will become more reliable and "always on," so they will be available at any time. PCs will be incorporated into everything, including cars, TVs, and phones.

Handheld devices are the story of the decade so far. Sales of PDAs and cell phones with Internet access have soared and are expected to rise from $10 billion in 2002 to $73 billion in 2005. It is not surprising that businesses and application developers are eager to enter the handheld market.

Personal digital assistants, or PDAs, combine a number of functions in a small, handheld PC. The Handspring Treo 180, introduced in 2002, is a cell phone, E-mail system, instant-messaging program, and limited Web browser. Future applications will expand the use of PDAs even more.

Handhelds, or pocket personal computers, weigh about one pound or less. They are used by package-delivery people to send hourly reports to headquarters. Car-rental representatives use them to process car returns from the middle of airport parking lots. Police officers access databases to check on suspicious car license numbers.

And now they are being used for more general business and personal tasks. For example, in January 2002, the Handspring Treo 180 was introduced. This Palm OS-based PDA is a cell phone, E-mail system, instant-messaging program, and limited

The Windows XP Media Center Edition, shown here running on Hewlett Packard Media Center hardware, is a home-entertainment system that hooks up a PC with a 42-inch, flat-panel TV display. Software and hardware companies are hoping the PC will become the command center for the high-tech home-entertainment center of the future.

Web browser all in one. The innovative mini-QWERTY keyboard lets users type e-mails and enter data with ease. One model even has a color display.

Many applications have already been rolled out for handhelds, and many more that are in development promise to have a bigger impact on corporate computing than the laptop computer ever did. "Handhelds are coming through the door," says senior IDC analyst Kevin Burden. "Companies are using them for their personal information management functions, but there's not a lot of value or return to the corporation if they're just going to be used for personal productivity. Only when corporate data is downloaded to these devices will corporations embrace them." [96]

Another entrant in the extended PC mix is the new breed of remote-controlled PCs that play DVDs and CDs and record TV programs. According to a report by Forrester Research, more than three of every five users already listen to music on their personal computer, while nearly half use it to watch DVDs. PC makers are betting that most home computers will no longer be confined to hidden-away spaces. They are hoping that the PC will take on a new role in the living room as the entertainment center.

A new home-entertainment system today features a PC with a Pentium 4 processor hooked up to a 42-inch, flat-panel TV display. It might also include additional memory, advanced video capabilities, a TV tuner, and a CD/DVD drive. The system runs a special version of Microsoft's operating system called Windows XP Media Center. It comes with a remote control that features a Start button just like the one in the desktop PC version of Windows.

The power and flexibility of PCs continue to escalate. In late 2001, Intel broke the two-gigahertz barrier for processor chips. "By 2030," says Bill Joy, Sun Microsystems' chief scientist, "we'll be building machines that are a million times more powerful than today's PCs." He pauses, then adds, "A million is a very big number." [97]

Stewart Brand, author and founder of The Well Community on the Web, thinks this is happening because computing has the inherent property of accelerating itself under the principle of Moore's Law. He says, "The reason Moore's Law keeps being true is that the first thing you do with each generation of denser chips is use them to make even denser chips. It's also the reason computer technology is the dominant, pace-setting technology that everything else is always sprinting to keep up with." [98]

The personal computer, once a truly revolutionary idea, has become a commonplace tool. It has brought computing power to the desktop and empowered people through their own fingertips. Changing the world is the PC's job.

1890 Hollerith's punched-card machine calculates census data.

1942 A working model of the ABC (Atanasoff Berry Computer) is completed.

1946 The ENIAC computer, the first fully electronic computer, is built.

1952 The UNIVAC computer predicts the election of President Eisenhower.

1954 The IBM 650 mainframe is offered for sale.

1963 The PDP-8 minicomputer is introduced.

1971 Intel introduces the first microprocessor, the 4004.

1974 Intel develops the 8080 microprocessor; the Mark-8 microcomputer is announced in *Radio Electronics* magazine.

1975 The Altair 8800 microcomputer is announced in *Popular Electronics*; Bill Gates and Paul Allen found Microsoft; the Homebrew Computer Club holds its first meeting.

1976 The Apple I microcomputer is introduced.

1977 The Apple II microcomputer is introduced.

1979 The VisiCalc spreadsheet and WordStar word-processing programs are released.

1981 The IBM PC is introduced.

1982 Lotus Corporation announces the release of Lotus 1-2-3.

1983 Compaq begins shipping its version of the PC.

1984 Dell Computer is formed; Apple introduces the Macintosh computer.

1985 Microsoft releases Windows 1.0.

1986 Compaq introduces a 386-based machine.

1990 Tim Berners-Lee invents the World Wide Web.

1992 Microsoft releases Windows 3.1; Michelangelo virus appears.

1993 PCs incorporate multimedia features.

1998 Apple releases the iMac; antitrust suit is filed against Microsoft.

1999 Intel introduces the Pentium III microprocessor; Concern grows about possible PC problems in Y2K (Year 2000).

2000 Microsoft unveils the Windows 2000 operating system; Intel introduces the Pentium 4 microprocessor.

2001 Dell Computer becomes the top seller of PCs; The PC market experiences negative growth for the first time in history. Microprocessor speeds break 2 GHz.

2002 The one-billionth PC ships; A federal judge approves the settlement of the antitrust case between Microsoft and the Justice Department.

2003 Tablet PCs become widely available, launching new era of mobile computing; sales of LCD flat-panel monitors exceed sales of CRT monitors.

1 Otto Friedrich, "Machine of the Year: The Computer Moves In," *Time*, January 3, 1983, 14–24. <http://ei.cs.vt.edu/~history/Time.MOTY.1982.html>

2 Phil Lawson and Robert Lindstrom, "A PC on Every Desk," *Sphericity*, 2000. <www.sphericity.com>

3 Quoted in Robert X. Cringely, *Accidental Empires* (Reading, Mass.: Addison-Wesley, 1992), 99.

4 Bill Gates, "Introduction," in *Inside Out: Microsoft in Our Own Words*, ed. Microsoft (New York: Warner Books, 2000).

5 Steven Lubar, *InfoCulture* (Boston: Houghton Mifflin, 1993), 10.

6 Alan Brightman, *Computers in Society*, ed. Karen Schellenberg. (Guilford, Conn.: Dushkin Publishing Group, 1992), 63.

7 Stephen Segaller, *Nerds 2.0.1*. (New York: TV Books, 1998),19.

8 Paul Freiberger and Michael Swaine, *Fire in the Valley* (New York: McGraw-Hill, 2000), xvi–xvii.

9 Friedrich, "Machine of the Year."

10 Quoted in Beatrice E. Garcia, "Looking Back at the PC 20 Years after IBM Put Its First One on Sale," *Miami Herald*, August 8, 2001.

11 Quoted in Larry Long and Nancy Long, *Computers* (Upper Saddle River, N.J.: Prentice Hall, 2001), 143.

12 Quoted in Segaller, *Nerds 2.0.1*, 140.

13 Quoted in M. Mitchell Waldrop, *The Dream Machine* (New York: Viking, 2001), 428.

14 Quoted in Freiberger and Swaine, *Fire in the Valley*, 46.

15 Quoted in Freiberger and Swaine, *Fire in the Valley*, 46.

16 Bill Gates, *The Road Ahead* (New York: Viking, 1995), 18.

17 Quoted in "The Computer Industry: Today and Yesterday" (from *Newsweek*, August 24, 1981). <www.microsoft.com/presspass/events/pc20/docs/TodayYesterdayFS.doc>

18 Cringely, *Accidental Empires*, 4.

19 Cringely, *Accidental Empires*, 4.

20 Quoted in Friedrich, "Machine of the Year."

21 Quoted in Friedrich, "Machine of the Year."

22 "Most Important Software Products," *Byte*, September 1995. <http://www.byte.com/art/9509/sec7/art5.htm>

23 Quoted in Mary Bellis, "The First Spreadsheet: VisiCalc: Dan Bricklin and Bob Frankston," *About.com*. <http://inventors.about.com/library/weekly/aa010199.htm>

24 Quoted in Lubar, *InfoCulture*, 346.

25 Friedrich, "Machine of the Year."

26 Quoted in Stephen H. Wildstrom, "The PC: Imperfect and Indispensable," *Business Week Online*, September 3, 2001.

27 Wildstrom, "PC," September 3, 2001.

28 Quoted in Freiberger and Swaine, *Fire in the Valley*, 239.

29 Nick Stam, "Just One More Test!" *PC Magazine*, March 12, 2002, 161.

30 Cringely, *Accidental Empires*, 14–15.

31 Cringely, *Accidental Empires*, 17.

32 Cringely, *Accidental Empires*, 16.

33 Cringely, *Accidental Empires*, 114.

34 Quoted in Cringely, *Accidental Empires*, 102.

35 Quoted in "We Told You So," *PC Magazine*, March 12, 2002, 165.

36 Freiberger and Swaine, *Fire in the Valley*, 380.

37 Quoted in Marc Ferranti, "Dataquest: Economy Curbs PC Recovery," *IDG.net*, November 15, 2002. <www.e-businessworld.com>

38 Quoted in Tobi Elkin, "PC Suffers Birthday Blues," *Advertising Age* 72, no. 35 (August, 27, 2001): 6.

39 John Heilemann, "Second Coming," *PC Magazine*, September 4, 2001.

40 Quoted in Heilemann, "Second Coming."

41 Friedrich, "Machine of the Year."

42 Clifford Stoll, *Silicon Snake Oil* (New York: Doubleday, 1995), 131.

43 Joseph M. McCade, "Technology Education and Computer Literacy," *Technology Teacher* 61, no. 2 (October 2001): 9.

44 Ronni Rosenberg, "Debunking Computer Literacy," in *Computers in Society*, ed. Kathryn Schellenberg. (Guilford, Conn.: Dushkin Publishing Group, 1992), 80.

45 "All Children Must Be Ready for a Different World," *International Society for Technology in Education*. <www.iste.org>

46 Quoted in Greg McDonald, "A PC on Every Desk," *Government Technology*, September 3, 2002. <www.stateline.org>

47 McDonald, "A PC on Every Desk."

48 Joe Celko, "PC Promises," *Intelligent Enterprise*, November 15, 2002.
49 Stoll, *Silicon Snake Oil*, 136.
50 Stoll, *Silicon Snake Oil*, 138.
51 Quoted in Fred O. Williams, "Buffalo, N.Y., Program Helps Meet Goals of Computer Literacy," *Buffalo News*, September 25, 2002.
52 Martin Schrader, "Learning to Live with Office Technology: A Senior Citizen's Story," *Westchester County Business Journal* 36, no. 5 (February 3, 1997): 23.
53 "Computer and Communications Use in Low-Income Communities: Models for the Neighborhood Transformation and Family Development Initiative," *Community Technology Centers' Network*. <www.ctcnet.org/casey/casey6.htm>
54 "Computer Use in Low-Income Communities." <www.ctcnet.org/casey/casey2.htm>
55 Donald A. Norman, *The Invisible Computer* (Cambridge, Mass.: MIT Press, 1998), viii.
56 Walter S. Mossberg, *The Wall Street Journal Book of Personal Technology* (New York: Times Books, 1995), 32.
57 Peter Schmideg, "It's Never Too Late," *Macworld* 19, no. 7 (July 2002): 34.
58 Quoted in Mossberg, *Personal Technology*, 37.
59 Norman, *The Invisible Computer*, x.
60 Reid Goldsborough, "'Migration' Anxiety: Moving from Old Computer to New," *New Orleans CityBusiness*, February 11, 2002, 19.
61 Howard Moscowitz, "Overcoming Computerphobia," *PC World*, December 1992, 35.
62 Florence King, "The Misanthrope's Corner," *National Review* 53, no. 7 (April 16, 2001): 64.
63 Quoted in Stacey C. Sawyer, Brian K. Williams, and Sarah E. Hutchinson, *Using Information Technology* (Chicago: Irwin, 1997), 307.
64 Quoted in Ben Gottesman, "Beyond the Backlash," *PC Magazine*, March 12, 2002, 102–103.
65 Quoted in Lubar, *Infoculture*, 347.
66 Quoted in "We Told You So," 164.
67 Ronald Harwin and Colin Haynes, *Healthy Computing* (New York: AMACOM, 1992), 2.
68 Harwin and Haynes, *Healthy Computing*, 33.
69 Janet Raloff, "Allergic to Computing?" *Science News* 158, no. 17 (October 21, 2000): 269.
70 Paul Marxhausen, "Computer Related Repetitive Strain Injury."
71 Quoted in Sally McGrane, "Creating a Generation of Slouchers," *New York Times on the Web*, January 4, 2001. <www.nytimes.com>
72 Quoted in McGrane, "Slouchers."
73 Quoted in McGrane, "Slouchers."
74 Friedrich, "Machine of the Year."
75 Levy, *Hackers*, ix.
76 Levy, *Hackers*, book cover.
77 Levy, *Hackers*, ix.
78 Quoted in Lubar, *InfoCulture*, 365.
79 Levy, *Hackers*, 26–33.
80 Quoted in Willie Schatz, "Operation: SunDevil," *The Washington Post*, Business Section, May 31, 1990.
81 Quoted in Schatz, "SunDevil."
82 Philip Elmer-Dewitt, "Bugs Bounty," *Time* 146, no. 17 (October 23, 1995): 86.
83 Christos J. P. Moschovitis, Hilary Poole, Tami Schuyler, and Theresa M. Senft, *History of the Internet* (Santa Barbara, Calif.: ABC-CLIO, 1999), 168.
84 Quoted in Segaller, *Nerds* 2.0.1., 274.
85 Quoted in Segaller, *Nerds* 2.0.1., 275.
86 Gates, *The Road Ahead*, 41.
87 Quoted in Segaller, *Nerds* 2.0.1., 266.
88 Quoted in Segaller, *Nerds* 2.0.1., 265.
89 Trixter, "Life Before Demos," September 25, 1996. <www.oldskool.org/shrines/lbd/>
90 Levy, Hackers, 377.
91 Trixter, "Life Before Demos."
92 Quoted in Freiberger and Swaine, *Fire in the Valley*, 379.
93 Ray Kurzweil, "Spiritual Machines: The Merging of Man and Machine," *Futurist* 33, no. 9 (November 1999): 16.
94 John C. Dvorak, "PC Paranoia," *PC Magazine*, September 1999, 87.
95 John C. Dvorak, "PC Paranoia."
96 Quoted in Michael Cohn, "Handhelds Get Down to Business," *Internet World* 6, no. 23 (December 1, 2000): 56.
97 Quoted in Heilemann, "Second Coming."
98 Quoted in Heilemann, "Second Coming."

"20 Years of Technology." *PC Magazine,* March 12, 2002.

Augarten, Stan. *Bit by Bit: An Illustrated History of Computers.* New York: Ticknor & Fields, 1984.

Bartholomew, Doug. "Lord of the Penguins." *Industry Week,* February 7, 2000.

Bowe, Frank G. *Personal Computers and Special Needs.* Berkeley, Calif.: Sybex, 1984.

Butler, Sharon J. *Conquering Carpal Tunnel Syndrome and other Repetitive Strain Injuries.* Berwyn, Penn.: Advanced Press, 1995.

Celko, Joe. "PC Promises." *Intelligent Enterprise,* November 15, 2002.

Cohn, Michael. "Handhelds Get Down to Business." *Internet World* 6, no. 23 (December 1, 2000).

Cringely, Robert X. *Accidental Empires.* Reading, Mass.: Addison-Wesley, 1992.

Dyson, Esther. *Release 2.0: A Design for Living in the Digital Age.* New York: Broadway, 1997. Elkin, Tobi. "PC Suffers Birthday Blues." *Advertising Age* 72, no. 35 (2001).

Freiberger, Paul, and Michael Swaine. *Fire in the Valley.* New York: McGraw-Hill, 2000.

Friedrich, Otto. "Machine of the Year: The Computer Moves In." *Time,* January 3, 1983.

Gates, Bill. *The Road Ahead.* New York: Viking, 1995.

Goldsborough, Reid. "'Migration' Anxiety: Moving from Old Computer to New." *New Orleans CityBusiness,* February 11, 2002.

Harwin, Ronald, and Colin Haynes. *Healthy Computing.* New York: AMACOM, 1992.

Heilemann, John. "Second Coming." *PC Magazine,* September 4, 2001.

Huff, Chuck, and Thomas Finholt. *Social Issues in Computing.* New York: McGraw-Hill, 1994.

Ifrah, Georges. *The Universal History of Computing.* New York: John Wiley and Sons, 2001.

King, Florence. "The Misanthrope's Corner." *National Review* 53, no. 7 (April 16, 2001).

Kurzweil, Ray. "Spiritual Machines: The Merging of Man and Machine," *Futurist* 33, no. 9 (November 1999).

Lawson, Phil, and Robert Lindstrom. "A PC on Every Desk." *Sphericity,* 2000. Available online at *http://www.sphericity.com*

Levy, Steven. *Hackers.* Garden City, N.Y.: Anchor Press/Doubleday, 1984.

Lubar, Steven. *InfoCulture.* Boston: Houghton Mifflin, 1993.

McDonald, Greg. "Maine Computer Project Engages Students." Stateline.org, September 3, 2002. Available online at *http://www.stateline.org*

McGrane, Sally. "Creating a Generation of Slouchers." *New York Times,* January 4, 2001.

Microsoft, ed. *Inside Out: Microsoft in Our Own Words.* New York: Warner Books, 2000.

Montgomery, Kate. *Carpal Tunnel Syndrome: Prevention and Treatment.* San Diego: Sports Touch, 1994.

Moschovitis, Christos J. P., Hilary Poole, Tami Schuyler, and Theresa M. Senft. *History of the Internet.* Santa Barbara, Calif.: ABC-CLIO, 1999.

Moscowitz, Howard. "Overcoming Computerphobia." *PC World,* December 1992.

Mossberg, Walter S. *The* Wall Street Journal *Book of Personal Technology.* New York: Times Books, 1995.

Norman, Donald A. *The Invisible Computer.* Cambridge, Mass.: MIT Press, 1998.

Raloff, Janet. "Allergic to Computing?" *Science News* 158, no. 17 (October 21, 2000).

Sawyer, Stacey C., Brian K. Williams, and Sarah E. Hutchinson. *Using Information Technology.* Chicago: Irwin, 1997.

Schellenberg, Kathryn, ed. *Computers in Society.* Guilford, Conn.: Dushkin, 1992.

Schmideg, Peter. "It's Never Too Late." *Macworld* 19, no. 7 (July 2002).

Schrader, Martin. "Learning to Live with Office Technology: A Senior Citizen's Story." *Westchester County Business Journal* 36, no. 5 (February 3, 1997).

Segaller, Stephen. *Nerds 2.0.1.* New York: TV Books, 1998.

Stoll, Clifford. *Silicon Snake Oil.* New York: Doubleday, 1995.

Sussman, Martin, and Ernest Loewenstein. *Total Health at the Computer.* Barrytown, N.J.: Station Hill Press, 1993.

Waldrop, M. Mitchell. *The Dream Machine.* New York: Viking, 2001.

WEBSITES

Byte.com
http://www.byte.com

Intel.com
http://www.intel.com

Inventors
http://inventors.about.com

"Chronology of Personal Computers"
http://www.islandnet.com/~kpolsson/comphist/

Microsoft.com
http://www.microsoft.com

PC World
http://www.pcworld.com

Freiberger, Paul, and Michael Swaine. *Fire in the Valley*. New York: McGraw-Hill, 2000.

Gates, Bill. *The Road Ahead*. New York: Viking, 1995.

Levy, Steven. *Hackers*. Garden City, N.Y.: Anchor Press/Doubleday, 1984.

Lubar, Steven. *InfoCulture*. Boston: Houghton Mifflin, 1993.

Segaller, Stephen. *Nerds 2.0.1*. New York: TV Books, 1998.

Stoll, Clifford. *Silicon Snake Oil*. New York: Doubleday, 1995.

Waldrop, M. Mitchell. *The Dream Machine*. New York: Viking, 2001.

WEBSITES

Byte.com
http://www.byte.com

Jones Media and Information Technology Encyclopedia
http://www.jonesencyclo.com

Intel.com
http://www.intel.com

Microsoft.com
http://www.microsoft.com

PC Magazine.com
http://www.pcmag.com

page:

Sandra Weber has an M.B.A from Temple University and a B.S. in mathematics and computer science from Clarkson University. She bought her first personal computer in 1983. After working for five years at Sperry Corporation as a technical writer and quality-assurance analyst, she became a computer science instructor. She now writes full-time about technology, history, and nature.